Written English

Written
English

A Guide for Electrical
and Electronic Students
and Engineers

Written English

A Guide for Electrical and Electronic Students and Engineers

Steve Hart

CRC Press
Taylor & Francis Group
Boca Raton   London   New York

CRC Press is an imprint of the
Taylor & Francis Group, an **informa** business

CRC Press
Taylor & Francis Group
6000 Broken Sound Parkway NW, Suite 300
Boca Raton, FL 33487-2742

Printed in Canada on acid-free paper
Version Date: 20150924

International Standard Book Number-13: 978-1-4987-3962-7 (Paperback)

Visit the Taylor & Francis Web site at
http://www.taylorandfrancis.com

and the CRC Press Web site at
http://www.crcpress.com

To ASEA, VDC, and JH.

Contents

Preface

Academics and students in the engineering schools of universities all over the world are required to write in English so their work can be included in international journals and be seen by as many peers as possible. Up to now there have been few resources available to guide these writers at the sentence level, to help them produce professional and accurate academic English and eradicate the errors made through bad habits or gaps in knowledge.

Unfortunately, a research paper let down by weak English is likely to be rejected by a journal editor and the author's credibility potentially questioned; likewise, the reputation of a student will be harmed if they fail to show sufficient competence in English, regardless of their actual subject knowledge.

This book has been written for

Professors
Lecturers
Research officers
Graduate students
Industry workers

This book covers the following fields:

Electrical Engineering
Electronic Engineering
Computer Engineering
Communication Systems

Consulting a typical English language learning textbook, heavy on both instructions and exercises, is time-consuming for researchers and students with deadlines and unsuitable for academics with a fair grasp of the language. An easily accessible and at-a-glance resource that can be utilized during the course of writing an essay or paper is the obvious solution. Errors made in papers authored by writers whose first language is not English are often easily fixed. The issue is one of awareness – and this can be achieved by identifying the mistakes and then providing instruction on how to correct them.

If writers are to eradicate the written English mistakes they have been making throughout their academic career and study, then they need to be recognized and catered for. This unique guide sets out to do just that.

The resource is divided into two main parts.

- Twenty four sections that cover key areas of grammar, writing style and formatting.
- An a–z section of the terms most commonly misused by writers. Elements include correct usage, example errors, related errors, definitions and clarification.

Although theory is covered in the guide and grammatical terms explained where necessary, the engineering writer's experiences and requirements are at the heart of this resource; and contrary to most guidebooks, analysis and reasoning play a supporting role to real-world examples showing the language being actively used and misused.

The guide's unique characteristic is this emphasis on real-world examples*; rather than the author listing errors that language learners are expected to make, the mistakes that are actually being made in research papers and essays are addressed and directly resolved. Some are surprising, others are expected, while a few could perhaps be considered careless – but without them being captured and exposed they will continue to be made.

---

* Where necessary, examples have been modified to preserve anonymity while retaining the nature of the error.

Q and A

Q – How can I search for errors that I don't know I am making?

A – This guide can provide advice on known errors and reveal unknown errors in a number of ways. Areas of weakness can be pursued in the relevant sections in the first part of the book, whereas specific terms can be accessed via the index or by scanning the a–z section. The book can be employed as a reference as you work and as a general reader from cover to cover. Both of these methods will reveal errors that you are currently making but were unaware of – and this awareness and knowledge of the correct form should prevent them being made in the future.

Q – How do I go about finding a particular term in the book?

A – You can search by topic in the first part of the book or consult the index containing all the sections and a–z entries in which the term can be found.

This guide does not attempt to cover every single area of English grammar; what it does do is focus on the areas relevant to the electrical and electronic engineering scholar, trainer and student. I have chosen to omit certain parts of speech, punctuation marks and language errors simply because they rarely feature in papers on engineering topics. The value of this guide is that it has been informed by over four hundred engineering papers written by international students and academics. I only include mistakes that have actually been made, not those that could be made or perhaps should be made. The areas, the theory and the errors I have included are therefore those most relevant and, crucially, those most likely to elevate the English level of a writer to that of a native speaker – the primary aim of the book.

Steve Hart

Author

Steve Hart has been editing and proofreading for international academics and graduate students of engineering since 2005. A former high school teacher with a background in sociolinguistics, he has written grammar guides for the Indian market and produced coursebooks for several academies in the United Kingdom. He is currently English Skills Coordinator at a higher education institution in Cambridge, England.

Section I

Grammar

CHAPTER 1

Nouns

Introduction

This section examines three key areas: the two types of nouns formed from verbs, the countability of nouns based on the concept of boundedness, and compound nouns and their associated rules.

Noun formation

Nouns that are formed from verbs can name a person or a device through the suffixes –or and –er. They can also name the activity taking place, often by taking the suffix –tion. Recognizing these noun endings can help with differentiating the two noun types and identifying them from the root verb.

VERB	NOUN	NOUN
	ACTOR/DEVICE	ACTIVITY/CONCEPT
amplify	amplifier	amplification
attenuate	attenuator	attenuation
communicate	communicator	communication
compress	compressor	compression
conduct	conductor	conduction/conductivity
convert	converter	conversion
detect	detector	detection
generate	generator	generation

(Continued)

VERB	NOUN	NOUN
	ACTOR/DEVICE	ACTIVITY/CONCEPT
identify	identifier	identification
induct	inductor	induction
interrupt	interrupter	interruption
manipulate	manipulator	manipulation
mediate	mediator	mediation
modulate	modulator	modulation
operate	operator	operation
oscillate	oscillator	oscillation
receive	receiver	reception
reflect	reflector	reflection
regulate	regulator	regulation
resist	resistor	resistance
respond	responder	response
simulate	simulator	simulation
subscribe	subscriber	subscription
transform	transformer	transformation
transmit	transmitter	transmission

Countable and uncountable nouns

Boundedness tells us whether a noun can be counted or not and therefore whether a plural can be formed. The concept is helpful in explaining why some nouns can be both countable and uncountable. To know if a noun can be counted we need to work out whether it has a clear 'boundary' and can be seen as a clearly separate thing, either physically or in our minds. Does it have a clear beginning and end?

Computer – the noun 'computer' is a separate entity that can be counted (**Computers** *were not able to carry out this task in the 1980s*).

Many plural nouns end in 's' or 'es' but there are a few different variations. Keep an eye on these particular plurals as they tend to be troublesome:

SINGULAR	PLURAL	PLURAL ERROR
analysis	**analyses**	analysis
antenna	**antennas**	antennae (mainly used for insects)
axis	**axes**	axi/axises

(Continued)

SINGULAR	PLURAL	PLURAL ERROR
diagnosis	**diagnoses**	diagnosis
flux	**fluxes**	fluxs
index	**indexes/indices**[a]	indexs
intermediary	**intermediaries**	intermediarys/intermediares
latency	**latencies**	latencys/latences
modulus	**moduli**	moduluses
stimulus	**stimuli**	stimulises

[a] Usually financial.

—data

Although used both as a singular and a plural, it is common practice to use data in the plural form – avoiding if possible the singular 'datum' by using a quantity term such as 'piece of'.

*The modulated data **are** coded in (2.3).   One (**piece of**) data that we have acquired…*

—uncountable nouns

Nouns that cannot be counted do not have clear boundaries. They have no clear parts that can be separated or enclosed. They are all part of the whole without any obvious limits. These uncountable nouns are usually concepts, abstract ideas, qualities, substances or emotions.

intelligence   safety   caution   equipment   information   evidence

Safety – the noun 'safety' cannot be thought of as having clear boundaries or limits.

*They must also guarantee **the safety** of all their engineers.*

If the noun is uncountable then it has no plural form.

~~Safeties~~ must also be taken into account when handling the cables.

Here are a few more:

access	entertainment	management	software
advice	hardware	patience	support
assistance	help	progress	tolerance
consumption	knowledge	reliability	traffic
coverage	literature	research	transport
electricity	luck	safety	trust

—uncountable and countable?

The confusing thing is that some nouns can be both countable and uncountable, depending on the way the noun is being used by the writer and also the context.

*The plan for the daily operations was initially drawn up on **paper**.*

*I had a number of **papers** that needed to be submitted before the end of term.*

The first example of the noun 'paper' is referring to the substance and is therefore uncountable. But in the second example the writer is referring to specific essay papers which can naturally be counted (they have a number of them).

Nouns that can be both countable and uncountable are often countable when the writer is referring to a specific instance or kind and uncountable when a general concept or sense is intended. Some nouns take on different meanings in their countable and uncountable forms (e.g. ground/grounds, power/powers).

absence	effect	level	strategy
achievement	environment	light	strength
assessment	evaluation	performance	success
behaviour	experience	perspective	teaching
classification	fire	policy	technology
communication	glass	power	theory
concern	government	prediction	thought
context	ground	pressure	time
control	growth	protection	travel
degree	industry	reality	understanding
development	influence	security	university
difficulty	language	society	work

Experience *plays a large part in being able to assess the requirements of the system.*

*We then surveyed the **experiences** of the users.*

Take care

When faced with a choice, many writers* will opt to use a plural when actually the uncountable (concept) form would be more suitable:

*This may well lead to ~~misunderstandings~~ (**misunderstanding**) and ~~conflicts~~ (**conflict**).*

In the above example the writer was not describing particular misunderstandings or conflicts, but writing in a more general sense. So when the uncountable form is required the context will relate to theory rather than to events.

All the network interfaces here will require **protection**.

The grid **protections** *are transmission line, busbar and backup.*

Writers often choose the plural form for the first example but this would be wrong.

----

* This book uses the term 'writers' to mean academics and students whose first language is not English.

Q and A

Q – How can a concept or idea have a countable form?

A – Most things you can see can be counted but there are also many things that you cannot see or touch that can be counted.

They have a number of **plans** that they are looking to implement.

The **hypotheses** will now be stated.

The uncountable form is the abstract or general sense and the countable is the concept actively being used in a situation.

Compounds

Compound nouns and phrases can be formed from combinations of nouns, adjectives, verbs and prepositions but they are always treated as a single unit.

*The next stage is to use this **network block**.*

When the term consists of more than one word, the first word acts like an adjective (it may even be an adjective) and modifies the second word.

*This interface will not permit any **user traffic** to…*

When the first word is a noun like this the temptation is to use the plural form.

*There would be no power flow through this line after ~~faults~~ (**fault**) isolation.*

Fault isolation is a compound noun made up of two nouns. The first word of the compound is always in singular form unless it has a plural only use, e.g. customs/earnings. The noun takes on a generic meaning here rather than detailing an actual situation.

Chapter 2

Articles

Introduction

An article is a word that is used before a noun to indicate the kind of reference being made to the noun.

the definite article = *the*   the indefinite article = *a/an*   the zero article

Articles in English help the reader or listener to identify and follow the nouns in a sentence and to understand the relationship between them and the other parts of the sentence. The terms in bold are where the writer has had to make an article decision for the noun 'algorithm':

> **Algorithms** *show the set of operations to be performed. In* **a cyclic algorithm** *the variables are partitioned.* **The cyclic algorithm** *used in this paper is a novel one and should lead to optimal values.* **Algorithm choice** *will depend on a number of factors...*

The definite article

If the writer can single out a noun as unique from the context of the sentence, and if the reader will understand the exact thing being described, then the noun can take a definite article.

So, for a noun to be definite the reader must be aware of the exact thing the writer is referring to.

*This algorithm turns the network into a **multilevel network** to substantially reduce computation time. Based on **the multilevel network**...*

> The definite article is used here because the writer has already introduced the network so the reader knows which network is being referred to.

Therefore, one way a noun can be definite is if it has already been mentioned.

A second way that a noun can be definite is the reader having prior knowledge of the noun because of a logical situation.

A research study looked at the performance robustness [2].
***The researcher's** findings implied that...*

> Here, the researcher had not been previously mentioned but the reader should be aware that a research study is carried out by a researcher. The reader can make an association between the two nouns.

The reader can also be made aware of the definiteness of the noun if the phrase following it creates a direct association based on the physical surroundings.

***The programmers** in the lab looked quite nervous so I went in and introduced myself.*

Similarly, the definite sense can be used in this next example because the reader will again understand the context. The reader will know that the clock being referred to is in the room where the presentation took place.

*After we presented our device, we looked at **the clock** and realised we had spoken for over twenty five minutes.*

The writer can also use a definite article when referring to somebody or something that there can only be one of. This is apparent in terms like *most, best, least* and *last*. These terms exclude all other things and leave just one – so logically the reader will understand that the noun is definite, that only one is being referred to and which one that is.

*For an inexperienced system analyst this is considered **the best** option.*

Rank (first, second, third…) can also be exclusive and make the noun definite.

*But on **the third** link there is no reply at all.*

The definite article can also be used when a singular noun represents its *entire type* in a general sense. The noun stands as an example of its type (like a prototype) and a general statement is made about this type as a whole.

***The engineer** plays a crucial role in this.*

> Here, the engineer is being used to represent all engineers. The writer is not referring to a specific one but engineers as a whole.

So the awareness of the reader can allow the writer to give nouns definite articles. But the writer must also ensure that the noun is able to be singled out for definiteness from the context of the sentence.

Changing the layout also had ~~the~~ significant effect on the results.

Here, the writer cannot single out 'significant effect' as unique and definite because it is not the only significant effect and this is additional information that the writer is simply using in a general way. The indefinite article (*a*) is required instead.

*Changing the layout also had **a** significant effect on the results.*

The indefinite and zero articles

A noun is used in an indefinite way when the writer either knows the reader will not be aware of the exact thing being mentioned or does not require the reader to know.

***A company** in the UK distributes these three phase auto-transformers. **Orion Ltd** has…*

In the sentence above the writer has yet to introduce the company to the reader so the noun is indefinite. They then go on to mention the company in the next sentence.

*Orion Ltd has found a way to fix this and so has **a company** in Japan.*

> In this sentence the writer has mentioned a company in Japan but only to make the reader aware that there is another company (as well as Orion Ltd) that has solved the problem. The reader does not know this company and the writer has no intention of mentioning the company again or giving any more details about them.

If the noun is singular and countable then an indefinite article ('a/an') will be used. If it is uncountable or plural then a zero article (no article) is most commonly used.

*They are looking for **government support** and also **funding** from **private investors**.*

> The exact type or kind of funding and support and who these investors are have yet to be identified. So the writer uses zero articles for the three instances and they have the equivalent meaning of 'some'.

The zero article can also be substituted for 'some' in the following example:

*But there will be (some) **companies** developing their own technology.*

The noun is also indefinite when it is being used as an example of its kind or type. The writer here is being generic and uses a singular countable noun:

A network user *must ensure that they have security in place.*

If a plural noun is used to achieve this then a zero article will be required.

Network users *must ensure that they have security in place.*

—specific does not necessarily mean a definite article

These last examples were generic, but it is important to understand that the noun can be indefinite even if the writer is using it in a specific sense. It is only ever definite if the reader has exact knowledge of the particular one or thing. This was seen in the earlier example '...*so has a company in Japan*'. The company exists (it is a specific company) but it has not been identified and the reader does not have sufficient information.

specific

*There was **a user** that left the group earlier than expected.*

indefinite

Abstract nouns relate to things that do not physically exist, in other words you cannot touch or see them. They tend to be emotions or ideas.

The writer here probably does not know the actual individual who left. All that matters is informing the reader that a user left.

A few uncountable nouns that relate to concepts or states (abstract nouns) can take an indefinite article in certain situations. If in doubt though just use 'type of' or 'kind of'.

*This is **a protection** that is not compatible with flow-like behaviour.*

Q and A

Q – When should I use 'an' instead of 'a'?

A – Use 'a' when the noun following has a consonant sound when it is spoken.

> *a device   a node   a circuit*

Use 'an' when the noun following has a vowel sound when it is spoken.

> *an expert   an approach*

If a noun beginning with 'h' is spoken softly like an 'o' we use 'an'.

> *an hour   an honest error*

If a noun beginning with 'eu' or 'u' has a 'you' sound then use 'a'.

> *a European research project   a unique approach*

If a noun beginning with 'o' has a 'wa' sound then 'a' is used.

> *a one-time process*

For numbers, remember to think of how it is spelled when written.

> *An 18-user system (eighteen)   An 11-node cluster (eleven)*

Also for abbreviations, think about the sound of the first letter.

> *an SIDR approach (S = es)   a BGP speaker   an MD5 code (M = em)*

Remember that the word immediately after the indefinite article will indicate whether you use 'a' or 'an'.

> *A signal has been recovered   An additional signal has been recovered*

—a typical article choice scenario

Every time a writer encounters a noun or noun phrase they must select an article to use. If we take the noun 'scenario' and assume the writer has recognised its countability, then there are several options.

- A definite article (*the scenario*)
- An indefinite article (*a scenario*)
- Plural form with a zero article (*scenarios*)
- Plural form with a definite article (*the scenarios*)
- Another determiner word (*this scenario/each scenario...*)

> 'Scenario' in the singular form can only take a zero article if it is part of a compound.
> **Scenario 2** *will involve...*

The only option the writer does not have and the one they often choose is a zero article with the singular form:

Of course, ~~scenario~~ will help us to analyse...

—communication

Communication is a good example of a noun that writers frequently use in a countable way when the uncountable form is more appropriate. Unless a particular act of communication is being referred to, the writer is likely to require the uncountable form with a zero article to signify a general meaning.

*The server runs a scheduler to initiate ~~communications~~ (**communication**).*

This reliance on the plural may be a consequence of many journal and article titles, and the field itself, adopting the plural form: *IEEE Trans. Wireless Communications.*

> *Communication is used in a general sense here so the plural (and hence the countable form) is not suitable.*

Proper nouns

A proper noun is different from a typical or common noun in that it states the actual name of the person, place or thing.

Common noun: *person*
Proper noun: **John Williams**

Common noun: *company*
Proper noun: **Orion Ltd**

Although they have a definite sense, proper nouns will usually take a zero article.

We would like to thank ~~the~~ Orion Ltd and ~~the~~ Dr Azman Reno for their valuable input.

However, proper nouns take a definite article if 'the' is seen as part of their name or common usage has led them to take 'the'.

*I will now compare the technology available in both **the UK** and Hong Kong.*

Fixed phrases

Fixed phrases are phrases that are in general use and are familiar to native speakers. Some of these phrases do not contain a definite article even though they have a definite sense. The meaning of these phrases may also be different to what would be expected. In other words, looking at the literal meaning of each individual word may not reveal the true meaning of the phrase.

*The archived websites are likely to be **out of date**.*   ~~out of the date~~

*This was covered **at length** in [5].*   ~~at a length~~

*The information will certainly be **of interest** given the situation.*
~~of their interest~~

—common errors

at first	**by accident**	**by design**
at ~~the~~ first	by ~~an~~ accident	by ~~a~~ design
by surprise	**from scratch**	**in advance**
by ~~a~~ surprise	from ~~the~~ scratch	in ~~an~~ advance
in place	**in private**	**in turn**
in ~~the~~ place	in ~~the~~ private	in ~~a~~ turn
on purpose	**on/in time**	**out of coverage**
on ~~the~~ purpose	in ~~a~~ time	out of ~~a~~ coverage
out of service	**take action**	**upon receipt**
out of ~~a~~ service	take ~~an~~ action	upon ~~a~~ receipt

*If this occurred then the network operator would have to ~~take an action~~ (**take action**).*

*We were hoping to find a procedure that simulates the BGP ~~from the scratch~~ (**from scratch**).*

Omitting the article

As scientists often look to describe general principles and processes, the definite article is disregarded at times by writers of technical papers.

Testing of this phase *was done over the course of three weeks with two trainees.*

Things being tested are often stripped of their definite status and made generic, despite the fact that specific events are being described. This is especially true of plural nouns.

*This involved **nodes** subscribing to **multicast groups** then **routers** registering traffic to **receivers**.*

This style (although concise) will sound imprecise, become tedious and eventually prove ambiguous for the reader if employed throughout the text. An article can add precision and aid understanding – and in some cases be essential to the meaning. Indeed, those who judge articles to be largely unnecessary and believe better clarity can be achieved by striking out every single one should examine the following muddled extract:

> *Proportionality coefficient depends on physical parameters of soil. Image obtained by concatenating A-scans recorded along survey line is called B-scan... how fast signal is attenuated depends on electrical conductivity of ground. Higher dielectric, slower radar wave moves through medium. On other hand, wet material will slow down radar signal. Resolution will be discussed in detail in following section.*

Example errors

> Singular countable nouns must have articles attached to them unless they are part of a compound.

~~Barrier~~ (*A/The barrier*) *is formed in the middle of* **the channel** *and its height modulated...*

> Other countable nouns that seem to produce article errors (i.e. the article is omitted) include: *amplifier, analyst, antenna, approach, component, concept, connection, coupler, domain, fault, mechanism, method, network, protocol, server, signal, spectrum, switch, timer.*

We use an artificial neural ~~networks~~ (**network**) *to carry out load forecasting.*

> Plurals are often mistakenly used with indefinite articles and vice versa.

They do have ~~the~~ (**a**) *basic understanding of* ~~the~~ (**an**) *isolated connection.*

> Overuse of the definite article is caused by a failure to appreciate when a noun is a unique instance and when it is being used in general terms or as an example.

A third set of balanced phasors was introduced by ~~the~~ [2].

*This could affect ~~an~~ (**a**) user's message count…*

*~~Authors~~ (**The authors**) gratefully acknowledge _____ for partially funding this research.*

We consider this the step 2 in process. ✗   *We consider this step 2 in the process.* ✓

the   *the*

In previous section, ~~the~~ two major problems were discovered that were present in both algorithms. The problems were firstly the fragmentations caused to tree and secondly the segmentation of the smaller regions. The methods applied to resolve problems together with the description of the integrated solution are described in following sections.

a   *the*

~~The~~ Matlab is ~~the~~ high-level language for technical computing produced by MathWorks. It is highly recommended for developing ~~the~~ signal and image processing algorithms, because of its powerful tools, ~~a~~ visualisation capabilities and functionality.

—chapters, figures…

A definite article will often be used when referring to a chapter, figure, equation, etc.

*In **the next chapter** I will evaluate the different protocols.*
***The equation** is as follows:*

The determiner 'this' can also be employed.

*In **this chapter** I will approach the subject from a new direction.*

But when writing the number of the chapter, figure or equation it is effectively being named and so changes to a proper noun with a zero article.

As shown in ~~the~~ chapter 3, this is not the case for these distributed systems.
~~The~~ Fig. 4. below shows the relationship between the two factors.

And when using next, previous, following, etc. a number is unnecessary.

This will be shown in the following chapter ~~6~~.

In the next chapter ~~5~~ we address...

The previous equation ~~3.4~~ is also ◄— *useful for this task.*

Brackets should be used here.

The previous equation (3.4)...
or simply
(3.4) is also useful...

Take care

Often a noun will have information attached to it that modifies it in some way. Usually it is an adjective or another noun. The writer may forget to use an article with the original noun because the article is not directly next to the noun and the descriptive word or phrase creates a distraction.

*This will be **a lengthy and complex process** that is likely to continue for some hours.*

The article here is relating to the noun 'process' which is a few words along.

*They are **an extremely integrated network sites** spread over many domains.*

The article here is relating to the noun 'sites' which is a few words along; but a plural should not take an indefinite article.

CHAPTER 3

Pronouns and quantifiers

Introduction

Pronouns and quantifiers come before the noun and modify it.

Pronouns: *it, your, we, their...*

Quantifiers: *all, any, some, several...*

Knowing whether to use them with a countable or uncountable noun and a singular or plural verb form can be a challenge for many writers.

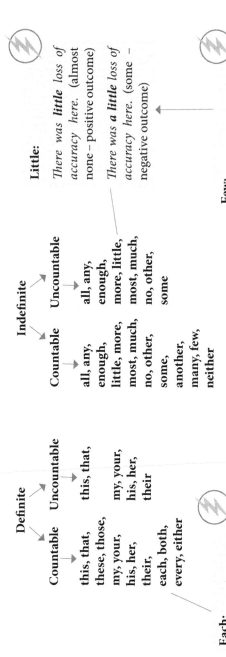

Definite

Countable

this, that,
these, those,
my, your,
his, her,
their,
each, both,
every, either

Uncountable

this, that,

my, your,
his, her,
their

Each:

These words are usually
used instead of articles:

~~The~~ *each method has
its merits as a potential
solution.*

Indefinite

Countable

all, any,
enough,
little, more,
most, much,
no, other,
some,
another,
many, few,
neither

Uncountable

all, any,
enough,
more, little,
most, much,
no, other,
some

Others like little, most and few can
be used like adjectives and can have
articles before them. But take note
of the difference in meaning with
and without an article.

Little:

*There was **little** loss of
accuracy here.* (almost
none – positive outcome)

*There was **a little** loss of
accuracy here.* (some –
negative outcome)

Few:

***Few studies** have looked at
this topic* (not many studies
– negative)

*A **few studies** have looked at
this topic* (some – positive)

When these words are the subject of the sentence it is always difficult to know whether to use a singular or plural verb. Some take singular, some plural and others both:

all – sing or plural	*each* – singular	*many* – plural	*several* – plural
another – singular	*either* – singular	*more* – sing or plural	*some* – sing or plural
any – sing or plural	*everybody* – singular	*most* – sing or plural	*something* – singular
anybody – singular	*everyone* – singular	*much* – singular	*that* – singular
anyone – singular	*everything* – singular	*neither* – singular	*these* – plural
anything – singular	*few* – plural	*other* – sing or plural	*this* – singular
both – plural	*little* – singular	*others* – plural	*those* – plural

Personal pronouns

The active voice is being encouraged a lot more in technical writing, and with it the first and third person pronouns for explaining how a process was carried out. It is also used for mathematics.

*In this paper **we** propose using a different parameter index...*

*Next, **we** define the vector as the optimal ESD.*

*In this framework, **they** study the problem under the assumption that all the packets are queued [10].*

—gender

To avoid gender bias the plural pronoun 'they' is increasingly being employed for the third person singular. 'They' is also used when the gender is unknown.

*The user can easily disconnect if he believes (**they believe**) this scenario has taken place.*

Of course, if a particular individual is being described then a gender specific pronoun is fine.

*The induction was given to the contractor before **he** entered the high voltage substation.*

'It' should not be used as a replacement for 'he/she' when referring to an individual.

*When the user encounters a problem it (**they**) will...*

*When the machine encounters a problem **it** will...*

—it

'It' should not be continually relied upon to refer to a previous subject because ambiguity can result; however, it is useful for avoiding repetition of the subject and acting as a general sentence starter.

'it' can be used in a general way where it does not refer to anything in particular but forms a general description or experience of what follows.

It is important to use the second switch.

Sometimes writers use 'this' when there is no direct reference with the preceding sentence. This is where 'it' can be employed as a link.

The signal strength can be weak in certain environments. Therefore, ~~this~~ (it) is essential for data integrity to be maintained.

But when used in this general way, 'it' cannot represent a noun and that noun cannot come afterwards.

It is important ~~the parameter~~ in the proposed scheme.

The sentence must be rearranged and the pronoun deleted.

The parameter is important…

A common error is using the possessive pronoun form (*its*) for a plural subject.

Modern interrupters were known for ~~its~~ (their) ability to produce rapid voltage surges…

The 'of' phrase

The phrase 'of the' can be used between a quantifier and a noun. By doing this the noun can take on a definite meaning.

Most of the people *think this is a good idea.* (definite: a specific group)

Most people *think this is a good idea.* (indefinite: people in general)

All of the networks *will struggle under these constraints.* (definite: a specific group)

All networks *will struggle under these constraints.* (indefinite: networks in general)

But/   ***None of the*** *schemes were a success.* ⟶ ***No*** *schemes were…*

Example errors

Each ~~messages~~ (**message**) is transmitted via this route.

> 'Each' is always singular so will have a singular noun and a singular form of the verb to be.

Each vendor can tune those parameters to suit ~~its~~ (**their**) own network conditions.

> Again, 'each' takes a singular noun but a vendor refers to a person not a machine or device so 'their' is required.

*Each transmitted ~~pulses~~ (**pulse**) has a particular time duration.*

*...and also has the ability to travel in both ~~direction~~ (**directions**).*

> 'Many' requires a plural noun

*Many of the key ~~parameter~~ (**parameters**) ~~shows~~ (**show**) a general variation.*

> The verb form also needs amending by dropping the 's' to reflect the plural.

*In both ~~model~~ (**models**) it is necessary to employ appropriate network pricing.*

*Several studies [11, 12, 21] ~~has~~ (**have**) shown that this leads to an effective solution.*

*This means all the ~~member~~ (**members**) of T1 ~~is~~ (**are**) mapped to only one member of T2.*

One of the problem(s) was calculating this route within the time allowed.

—all

Sometimes 'all' begins a sentence that is negative in nature. A better construction is to begin with 'no/none'.

All the users do not operate this device regularly.   ✗

None of the users operate this device regularly.   ✓

Another issue is when the writer actually means 'only some'. For this, 'not all' can be used.

All of the services do not run in synchronization.   ✗

Not all of the services run in synchronization.   ✓

For negation use 'any' not 'all' to mean none. In the sentence below 'all' would imply that it still worked on some of the systems.

*This does not work on ~~all~~ (**any**) of the systems.*

—another/other

For countable nouns use the terms in the following way:

another – singular (one other; a further)   other – plural (some other; further)

*~~Other~~ (**Another**) study **has** focused on the shortcomings of the traditional hierarchical...*

*~~Another~~ (**Other**) studies **have** focused on the shortcomings of...*

CHAPTER 4

Subject/verb agreement

Introduction

The main purpose of the three experiments ~~were~~ to identify the reason for the failure ✗	The **main purpose** of the three experiments **was** to identify the reason for the failure. ✓

In order to make the verb agree with the subject of a sentence, ask three questions:

What is the **subject** of the sentence?	*The **main purpose** of the three*
Is the subject singular or plural?	singular (*main purpose*)
Does the verb match the subject?	*was* (singular) YES

Making the subject agree with the verb is not a straightforward procedure. Here are some useful pointers to make the task easier:

To ensure that you select the correct verb form you must identify the subject.

The subject will be a person or place, idea or thing that is doing something or being something.

| singular subject | singular verb |

The user of this network   *is*   *likely to have poorer service.*

...but sometimes it is not easy to identify the subject

There may be a singular or plural noun in the sentence causing confusion. If it is not the subject it should not affect the verb form.

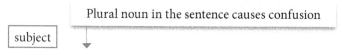

The **role** *of the amplifiers* ~~are~~ (**is**) *crucial here.*

Similarly with a singular noun,

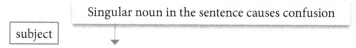

The **key issues** *for the engineer* ~~is~~ (**are**) *divided into four (see Table 11).*

The subject normally comes before the verb, but with sentences that are questions or begin with 'There' or 'It is' the verb can be found before the subject.

| verb | subject |

How   **do**   **you**   *measure the current in this instance?*

| verb | subject |

There **were** *three* ***main reasons*** *for the power loss:*

The subject may be a verb form ending in -ing.

| subject |

Researching *is the most useful way to solve these power issues.*

The subject may also be 'it' referring to a later expression:

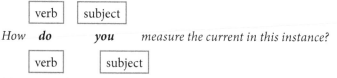

It *was apparent to everyone* ***that the antenna system needed changing.***

Using two subjects

Check that the subject has not been repeated in the sentence. Here a pronoun has been added unnecessarily.

| subject | subject repeated |

The **study** ~~it~~ *looks at a fixed-network scenario.*

| subject | subject repeated |

...and the **problems** ~~they~~ *were increased by poor development of this protocol.*

Singular or plural form?

The presence of errors | was/were | *more important to the success of the project.*

> In this instance 'presence' is the key phrase (not 'errors') so the singular form '**was**' is used.

> In this instance 'design goal' is the key phrase (not the 'projects') so the singular form '**is**' is used.

Our eventual design goal for these projects | is/are | *to create a user-friendly system.*

> In this instance 'looking' is the key phrase so the singular form '**is**' is used.

Thus, looking for suitable networks | is/are | *the next stage of the process.*

Words of quantity

When using phrases that indicate portions or quantity (majority, percentage, some...), the noun after 'of' will determine whether to use a singular or plural verb.

A third of the people are unhappy with the service. (*people* = plural)

A third of the country has no access. (*country* = singular)

Uncountable nouns

If the two nouns are separated by 'and' use a plural verb form.

– uncountable AND uncountable = plural

*Confidence and expertise **are** required for this.*

– uncountable OR uncountable = singular

*Confidence or expertise **is** required for this.*

However, some nouns together are considered a single entity and therefore take the singular verb form:

*Research and innovation **is** crucial to success.*

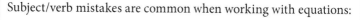

Take care

Subject/verb mistakes are common when working with equations:

*…where i ~~represent~~ (**represents**)…*

*…where i and j ~~represents~~ (**represent**)…*

In this next example the writer has been influenced by the two control signals and uses a plural form of the verb. In fact the subject is 'the relationship' and requires a singular verb:

*Rearranging (2–21), the relationship between the control signals M2d and M2q ~~are~~ (**is**):*

Mistakes also seem to occur with abbreviations:

*Our GMs also ~~shares~~ (**share**) the same SA. The DVM ~~were~~ (**was**) tested for…*

CHAPTER 5

Verbals

Introduction

Verbals represent a challenging area for the writer of English in that although they are derived from verbs, they actually function as nouns, adjectives or adverbs in a sentence.

Three types are recognised:

Gerunds – ***Finalizing**** the scenario is the next task.

> These verb forms end in –ing and act like a noun.

Participles – *We will then distribute the **routing** information.*

> These verb forms usually end in –ing or –ed and act like adjectives.

Infinitives – *….and then attempt **to define** the demand.*

> These verb forms are usually preceded by 'to' and act like nouns, adjectives or adverbs.

Choosing the –ing form*

It is useful to recognise the three different forms in your writing but the dilemma at the practical level is usually whether to use the 'to' form or the –ing form.

---

* The term –ing form will be used throughout to aid understanding.

It can depend on the verb that precedes the verbal. Some verbs will only be followed by the –ing form while others may need to be followed by the 'to' form.

The verb 'to suggest' will always be followed by the –ing form:

They suggest to change to a smaller transistor.   ✗

*They **suggest changing** to a smaller transistor.*   ✓

Other verbs that are followed by the –ing form include: avoid, consider, delay, finish, keep, postpone, recommend, require, risk.

*The method avoids **using** an unauthenticated channel.*

*We will have finished **measuring** the coherent bandwidth by then.*

A useful rule to remember is that the –ing form must always be used after a preposition.

*These methodologies are necessary **for improving** tolerance.*

Present and past participles

There are two kinds of participle, the present and the past. The present participle always ends in –ing.

*We are looking to reduce the **processing** time.*

The past participle ends in –ed for regular verbs but has various endings for irregular verbs.

*These **connected** elements then form a new cluster.*

These participles are used as adjectives and also help form the present and progressive tenses. Writers should consider the following two areas of difficulty:

—simple past and past participle

The past participle is used after the verb 'to have' to form the perfect tense and for passive constructions. If the verb is an irregular verb it will have a different form to the usual past tense form.

arise – irregular verb

*This problem **arose** at the end of the first stage.*   (simple past tense)

*...then identify the circuit in which the problem had **arisen**.*

> past perfect tense.
> past participle form is used.

It is a common error to use the simple past for an irregular verb in the perfect tense and in a passive sentence.

*We have ~~chose~~ (**chosen**) a high frequency region for our simulation.*

*A simpler model was ~~chose~~ (**chosen**) to aid understanding.*

—dangling modifiers

As well as a single participle modifying a noun as seen above, participles can also be found in phrases that collectively modify the noun.

***Handling the initial data**, the center will identify the patterns before the engineers...*

The phrase is modifying the subject of the sentence (*the center*) and appears directly before it. A common error is to create a dangling modifier whereby the phrase has no obvious subject attached to it or it modifies the wrong subject.

Addressing these issues, the software ran on a different setting...   ✗

Addressing these issues, we ran the software on a different setting...   ✓

The dangling modifier in the first example creates the impression that the software addressed the issues itself.

Choosing the infinitive

The verb 'to expect' will always be followed by the infinitive form.

We expect ~~seeing~~ the traffic rate increase considerably.   ✗

*We expect **to see** the traffic rate increase considerably.*   ✓

Other verbs that are followed by the infinitive include: agree, attempt, decide, intend, learn, need, plan, propose, want.

*We plan **to move** this up one more level.*

*This determines how many terminals want **to work** on the job.*

Some verbs can be followed by the infinitive form or the –ing form with little or no change in meaning: begin, continue, like, prefer, remember, start, try.

*The user will continue **switching/to switch** channels in this scenario.*

—verb+prep combinations

Some words have developed partnerships with particular prepositions, especially for referring to activities or outcomes. The verb that follows them must be in the –ing form (because of the preposition).

There is a tendency for writers to select the 'to+verb' form every time they are faced with this apparent 'choice'. Again, check the correct usage if unsure.

*capable of – They are capable ~~to form~~ (**of forming**) connections with these nodes…*

*succeed in – The question is whether this intervention will succeed ~~to improve~~ (**in improving**) the reliability of the system.*

Q and A

Q – Isn't sometimes the 'to' missing from the infinitive form?

A – Yes, this is most commonly seen after modal verbs and after an object when the main verb is hear, see, make or let.

A very common error is to retain the 'to' when these verbs feature.

Energy costs can be saved by letting nodes ~~to~~ stay idle.

We can then make the program ~~to~~ check the current situation.

Selected examples

Here are some common dilemmas and their solutions:

It is not just a matter of prevent/preventing an attack...

The –ing form is required because the preceding word is a preposition (of).

This mistake is often seen when 'before', 'after' or 'since' begin a sentence:

*Before ~~measure~~ (**measuring**) the time it takes to...*

This requires ~~to embed~~/embedding a mechanism to notify the controller.

The verb 'to require' necessitates that the verbal following must be in the –ing form.

They aim to raise/~~to raising~~ awareness of...

The endings of the infinitive form cannot change. It retains the dictionary form or plain form of the verb and is unaffected by tense or plurality. Here is another example:

*We hope ~~to evaluates~~ (**to evaluate**) these measures at a later date.*

~~Change~~/Changing the antenna might also work.

Here we need the –ing form at the start as the subject of the sentence.

*The engineer could then decide (**to**) sample the signals.*

Remember to include the 'to' part

Take care

So far we have looked at sentences where the verbal follows the main verb with no other words in between:

*It **continued to produce** the best results.*

Some verbs require an actor (a pronoun or a noun) between the main verb and the verbal.

A verb that always causes problems is 'to allow'.

This will allow to calculate the operation and maintenance costs.   ✗

This will allow (us/them/the regulator) to calculate the operation and maintenance costs.   ✓

Another option is to make the sentence passive: *This will allow... to be calculated.*

Other verbs that require an 'actor' are: advise, convince, enable, encourage, instruct, permit.

*We can then convince **the infectors** to forward the traffic...*

*The operator instructs **one of the neighbors** to block the user.*

CHAPTER 6

The verb 'to be'

Introduction

1. To take place; to occur
2. To exist or live

Present participle: **being**
Past participle: **been**

PAST		Simple		Perfect
	I	**was**	I	**had been**
	It	**was**	It	**had been**
	We/They	**were**	We/They	**had been**
		Progressive		Perfect progressive
	I	**was being**	I	**had been being**
	It	**was being**	It	**had been being**
	We/They	**was being**	We/They	**had been being**

PRESENT		Simple		Perfect
	I	**am**	I	**have been**
	It	**is**	It	**has been**
	We/They	**are**	We/They	**have been**
		Progressive		Perfect progressive
	I	**am being**	I	**have been being**
	It	**is being**	It	**has been being**
	We/They	**are being**	We/They	**have been being**

FUTURE	Simple	Perfect
	I **will be**	I **will have been**
	It **will be**	It **will have been**
	We/They **will be**	We/They **will have been**
	Progressive	Perfect progressive
	I **will be being**	I **will have been being**
	It **will be being**	It **will have been being**
	We/They **will be being**	We/They **will have been being**

Confusion with 'being'

The verb form 'being' is the present participle of the verb 'to be' and is used for the present progressive tense.

*We will now look at how they are ~~been~~ (**being**) managed.*

The form 'been' is the past participle but unlike most participles these two forms cannot be used as adjectives.

Sometimes 'being' and 'been' are confused but from the verb forms above it is clear that 'been' is always found after the verb 'to have'.

have been

There ~~been~~ issues with achieving simultaneous delivery.

been

Morgan et al. have ~~being~~ searching for a solution since 2005.

Errors also occur when the sentence includes

- Despite
- As well as/also
- Due to

The simple form of the verb (is/are/was/were) cannot be used in the following sentences, they require 'being'.

Despite these results ~~are~~ rather mixed, we can still draw a few conclusions.

As well as this ~~was~~ the simplest solution, it was also the most feasible.

'being' can also be a gerund.

> *My **being** involved in this project has given me the opportunity to increase my knowledge.*

And remember 'being' is the form to use after a preposition:

*There was also a problem **with being** too cautious about getting captured.*

Confusion between has been/was

Looking at the list in the introduction we can see that 'was' is the simple past form of the verb and 'has been' is the present perfect.

So 'was' is used to describe something that happened in the past and has now finished.

*He **was** also the chief engineer at the time [22].*

'has been' is used to describe something that happened in the past but the actual time of the event is not important. It may be linked with something continuing today. There is some overlap between 'was' and 'has been' but the error is made when a particular point in time is used.

A particular point in time means that the present perfect (*has been*) cannot be used.

There ~~has been~~ a re-issue in 1999.

*There **was** a re-issue in 1999.*

But 'has been' is used along with 'since' for a point in time if the event is still continuing today.

*The principal developer (who since 2011 ~~was~~ (**has been**) Antonio Durant) was also involved.*

Remember to include the 'been' part:

Many results have obtained for this particular frequency. ✗

Take care

Missing 'is'

Writers are prone to missing out 'is' in the sentence, especially when 'which' is involved.

*Orion, which (**is**) the largest network in this region...*

*The received signal of this subcarrier (**is**) expressed as:*

CHAPTER 7

Modal verbs

Introduction

Modals are auxiliary verbs that change the manner of a sentence and show the likelihood or ability of something. They are followed by a main verb and give extra information related to ability, possibility, necessity or willingness.

> can/could   will/would   may/might/must   should/shall

The verb form that follows is always the infinitive but without the 'to'.

*The grid **can provide** an additional link to connect other terminals.*

Golden rule

The golden rule to follow when using modal verbs is that the verb immediately after the modal retains its base or dictionary form.

*The program **begins** by initiating a request.   The program will **begin** by initiating...*

So even when the subject is singular, the verb can take the same form as the plural 'they' or 'we' when a modal verb is present.

***They improve** the quality of the service.   An engineer **will improve** the quality...*

Because the verb never changes, the –ing form cannot be used after a modal.

*Bandwidth restrictions can ~~causing~~ (**cause**) significant delays in receiving…*

—will/would

'Would' is incorrectly used by writers specifying what they plan to discuss in their work. Because of its role in conditional sentences, it sounds to the reader as though this was not actually possible.

> The only time 'would' can be used in this context is to express hope or justification. *This study would be able to solve this problem.*

We would discuss this in section four. The reader is waiting for the second part….

*We would discuss this in section four **but there is a lack of evidence**.*

Instead, 'will' informs the reader what follows because the future tense is required.

*We **will** discuss this in section four.   This **will** be covered in a later chapter.*

One issue that occurs with 'will' is the insertion of a past tense verb form.

*It will ~~amplified~~ (**amplify**) the high frequency component…*

—can/could

Use **can** to describe something in the present or future that is likely or certain to happen or that you are able to do.

We can determine how long this will take by applying the curve…

Use **can** for seeking or granting permission.

They can leave the group if they contact the system administrator.

Use **could** to describe something that is possible in the present or future, but may not happen for various reasons.

We could monitor each node but are restricted by time.

Use **could** to express alternatives.

They could also work on improving the design first and then testing it again.

'could' is the past tense of 'can' so is also used for past events.

They could access these files when they were guest users.

'Could' is often unnecessarily added in the present tense.

We ~~could~~ notice that the highest power of x that occurs…

Questions

Questions contain auxiliary verbs ('be', 'do', 'have') or/and modals. Sometimes these verbs are missed out in error.

~~They use tags for retrieval?~~   Can they use tags for retrieval?

In questions the modal verb comes before the subject:

***Can they** support multiple antennas?*

In a statement the subject comes first:

***They can** provide a wide variety of tasks.*

Take care

Modal verbs can be changed to negatives by adding 'not'. This is placed between the modal verb and the main verb. In informal writing negatives can be shortened by the use of an apostrophe, but this is not recommended for academic writing.

  could not (couldn't)   will not (won't)   should not (shouldn't)

*The grid ~~won't~~ (**will not**) operate normally during the fault.*

CHAPTER 8

Phrasal verbs

Introduction

Phrasal verbs are multi-part verbs made up of a verb and a preposition or particle. They are different to other phrases that contain prepositions in that the meaning is not obvious if the parts of the phrasal verb are considered separately.

*We must **point out** that the signals are initially sampled at low IF.*

These verbs end in directional words such as 'on', 'down', 'out' and 'back' but they are being used in an abstract way so judging them can be difficult. Many phrasal verbs are considered colloquial or examples of informal language so single-word verbs are preferable (e.g. start off – begin); but some have an important part to play in technical writing and are certainly useful as descriptive terms once their meaning has been acquired and use mastered.

A few examples

act on/upon	add on	back down	back up	break away
break down	break through	break up	build up	call back
call up	cancel out	carry forward	charge up	close down
come across	come back	come up against	come up with	cover up
cut off	cut out	depend on	drown out	ease off
fall back on	feed off	hack into	help out	key in
light up	line up	link up	lock out	log in/into/out
log on/off	look into	pass by	pass on	phase out
power up	result in	scale down	send back	set up
shut down	shut off	sign in/out	step down	take down
team up	tell apart	time out	track down	tune in to
tune up	turn on/off	turn up/down	type in	use up
wire up	zoom in/out			

Position of the particle/preposition

The position of the second word in some of these verbs is fixed, coming directly after the verb and before the object.

*This also **came from** the 40 RBs/subframe.*

In other verbs the second word can come either before or after the object.

*The user can then **print out** the code.*

*The user can then **print** the code **out**.*

Types of error
—confusion with single-word verbs

Some writers confuse two-word verbs with single-word verbs that are similar in meaning. Other writers overlook the single-word verb, which, if available, should always be chosen over a phrasal.

To ensure this does not ~~spread out~~ (**spread**), *new messages should be generated.*

It would be useful to ~~find out~~ (**discover**) *why this event took place.*

Unfortunately this further ~~held up~~ (**delayed**) *the procedure.*

Similar URLs have also been ~~left out~~ (**omitted**).

We ~~tried out~~ (**tested**) *our program in an unchanging network environment.*

—wrong particle

We first ~~focus in~~ (**focus on**) *circuit-level solutions…*

The devices must all be ~~switched of~~ (**switched off**) *at this point.*

They would then be able to ~~ease away~~ (**ease off**) *on the frequencies that were known to be less important.*

This increases the risk of them hacking ~~onto~~ (**into**) *the system.*

We predict that this will ~~result to~~ (**result in**) *a significant gain.*

Much will ~~depend with~~ (**depend on**) *the particular constraint limit.*

—confusion between particles

Some phrasal verbs have the same verb but a different particle and therefore a different meaning.

Statistical work has been ~~carried on~~ (**carried out**) *with this in mind.*

They would then have the ability to ~~take away~~ (**take down**) *this particular botnet.*

The user here ~~came up with~~ (**came up against**) *a more competent attacker.*

CHAPTER 9

Adjectives and adverbs

Introduction

Adjectives only ever modify nouns and pronouns. Adverbs are more flexible and can provide information about verbs, adjectives and other adverbs. The most likely decision facing a writer in this area is whether to use the adjective or the related adverb form ending in –ly (note that not all adverbs end in –ly).

This has been temporary/temporarily used to determine the domain type.

The word is modifying the verb 'use' and so an adverb will be required – temporarily.

Adjective/adverb comparison

Some similar looking adjectives and adverbs can have very different meanings:

economic (in the economy) adj	v	**economical** (money-saving) adj
They would need to look at the **economic** conditions at the time.		This method is more **economical** overall.

fair (sufficient; right) adj   v   **fairly** (quite) adv
This mechanism provides a   The performance estimate is
fair allocation of bandwidth   **fairly** accurate.

free (without paying) adj   v   **freely** (without restriction) adv
Users would not expect to   System engineers can **freely**
receive this for **free**.   access the source code.

high (great) adj   v   **highly** (very) adv
These spam filters have   A **highly** flexible platform was
high recognition rates.   created for this purpose.

late (arriving after the   v   **lately** (recently) adv
expected time) adj   **Lately**, more research has
Inconsistency resulted from   been carried out in this area.
this **late** response.

short (not long) adj   v   **shortly** (soon) adv
This includes a **short**   These factors will be
production time.   considered **shortly.**

ADJECTIVES FOR SHAPES

circular   cylindrical   add –ly to form the adverb and –ity for
                           the noun form:

linear   orthogonal

rectangular triangular   *The orthogonality of extended sequences*

Q and A

Q – If adverbs modify verbs how can adjectives come after verbs?

A – Adjectives that come after the verb are modifying the subject of the sentence, not the verb.

They do not describe the verb. That is the role of adverbs:

*The results were **disappointing.***

These adjectives often occur with the verb 'to be' and other linking verbs (e.g. become, feel, seem).

*The utility companies were **anxious** about this.*

They usually describe feelings.   *The users felt **glad** about this decision.*

Adverb placement

Adverbs are most effective when appearing before the modified word and have to be placed there when that word is an adjective:

*We have to ensure that the second route was **suitably** different.*

But adverbs that describe how something is done, adverbs of time and adverbial phrases can be found after the modified word and at the end of a sentence.

*We have to ensure that the video frames change **slowly**…*

*The operator would then forward these **at a later date**.*

As a rule, adverbs should be placed as close to the word being modified as possible to avoid confusion and ambiguity. The versatility (and also dangers) of adverb and adjective position can be illustrated in the following scenario.

Here the writer wishes to express that only the network manager can access the data, nobody else, so 3 is required. If the network manager had access to data but to nothing else then 2 would make sense but still sounds awkward. 1 is ambiguous and 4 is stylistic and would be more appropriate as part of a set of guidelines or as a warning message. To achieve the intended meaning, 'only' needs to be as close to 'available' as possible.

1. *The **only** data is **available** to the network manager*
2. *The data **only** is **available** to the network manager.*
3. *The data is **only available** to the network manager.*
4. *The data is **available** to the network manager **only**.*

Comparatives and superlatives

Adjectives and adverbs can be used to compare and there are three degrees of comparison:

attribute degree –	high	suitable (+)	suitable (–)	good
comparative degree –	higher	more suitable	less suitable	better
superlative degree –	highest	most suitable	least suitable	best

The comparative degree compares two things and the superlative degree compares at least three. Most adjectives and adverbs take on the forms above to move through the comparative stages. Notice that 'good' has an irregular construction and different words are used to create the comparison.

*This has **high** priority.*

*This has **higher** priority than the previous task.*

*This has the **highest** priority of all the tasks.*

*This proved to have the ~~lower~~ (**lowest**) power of all of the systems tested.*

Some adjectives cannot form these different degrees of comparison by changing their endings, so they use more/less and most/least instead. A common error is failing to recognize the ones that can.

*The node chooses the ~~most close~~ (**closest**) clusterhead with whom it shares a key.*

Note also how 'than' is used to complete the comparative form. Adjectives in the first degree do not technically compare so cannot be used with 'than'.

*Using this system the signal is **unreliable** ~~than~~ the strong signal achieved by the previous method.*

So if there is not a comparing word in the sentence 'than' should not be used. 'Compared with' can be used to form a comparison instead.

*Using this system the signal is unreliable **compared with** the strong signal achieved by the previous method.*

When there is an apparent choice between the two terms (than/compared with), opt for 'than' if the adjective is in the comparative degree and the two things are being evaluated directly.

*The fault current was also much greater ~~compared with~~ (**than**) the relay setting current.*

It is important to consider whether the reader will be able to identify what is being compared. It might seem obvious to the writer but has the comparison actually been formed?

The terminal has a lower initial current because of its higher cable resistance.

Lower than what?
Has another terminal been mentioned?

Higher than another terminal?
Higher than normal cable resistance?

The past participle

Present and past participles can function as adjectives. Past participles that end in –ed (most of them do) are often unwittingly written without the 'd' when paired with nouns.

*We will also demonstrate three types of ~~balance~~ (**balanced**) faults.*

*There were a number of issues with the ~~estimate~~ (**estimated**) error.*

Excessive adjectives

Adjectives should be used thoughtfully and only for description or emphasis. Do not use robust adjectives to persuade the reader of something or overemphasize a situation. Strong adjectives usually sound unprofessional, inappropriate and give the impression that the writer is trying too hard to convince the reader. Use more modest adjectives and phrases that have clear and direct meanings and that the reader will be more familiar with and will more likely accept.

*~~An incredible~~ (**A key**) observation is that the signal can be...*

*This is ~~of magnificent importance~~ (**crucial**) to the success of the scheme.*

*Through ~~immeasurable~~ (**much**) trial and error...*

USE	AVOID
crucial/critical/vital/key	absolutely indispensable/monumental
excessive	massive/giant
modest	tiny/miniscule
unusual/uncommon	remarkable/weird

These words do not require 'very' or 'extremely' before them.

Example errors

—small/few/little

'*little*' can precede uncountable nouns when referring to amount.

Little evidence *was found to support this view.*

'*few*' can precede plural nouns when referring to number.

Few people are actually aware of the need to back up this data.

'*small*' cannot go directly next to the noun when referring to amount or number.

~~Small evidence~~ has been collected already.

Only when it is relating to the actual size of something can it go next to the noun.

Small devices are more convenient and offer...

—good/well

good – adjective   well – adverb (can also be an adjective meaning 'healthy')

'good' is mainly used with nouns and comes before the noun.

*This policy strikes a ~~well~~ (**good**) balance between reliability and energy cost.*

So if you are describing a noun use 'good'. 'well' often describes a verb or adjective.

Also see A–Z   well/good

*They performed the task ~~good~~ (**well**).*

*Some of these policies have been **well developed**.*

—less/fewer	—such
less – not as much	Never follow 'such' with a number.
fewer – not as many	*~~Such one~~ example is...*
Use fewer for countable nouns:	Use an indefinite article if the noun is singular and countable.
*There are ~~less~~ (**fewer**) design constraints...*	*Such an example is useful for explaining this.*
Use less with uncountable nouns:	But not if the noun is plural or uncountable:
*We noted **less** evidence of this.*	*Such evidence will help to explain why...*
	Used with a noun in this way 'such' means 'this/that type of'.

Switches 1 and 2 are used to increase the load ~~by~~ manually.

> Often the preposition 'by' is mistakenly placed before the adverb. This would be OK if the sentence was to continue,
>
> *...to increase the load, by manually altering the...*

*From Fig.1.6, it is ~~clearly~~ (**clear**) that the structure repeats itself.*

*Many services run ~~simultaneous~~ (**simultaneously**), so the size of the log files will...*

*This can also help to coordinate various ~~node~~ (**nodes**).*

> 'various' modifies a plural countable noun not a singular countable.

*These support vectors are the samples furthest ~~to~~ (**from**) the hyperplanes*

> 'closest to', 'nearest to' but 'furthest from'.

The probability that an attacker succeeds is inverse proportion to j.   ✗

The probability that an attacker succeeds is inversely proportional to j.   ✓

*This provides a high data rate at a low cost with ~~relative~~ (**relatively**) low power consumption.*

*The third algorithm performed twice as fast ~~than~~ (**as**) the second one.*

> This sentence uses an adjective in the first degree (fast) so 'than' cannot be used. However, a comparative construction is possible by using the following form:
>
> *as+adjective+as*

CHAPTER 10

Prepositions

Introduction

There are few general rules relating to preposition choice; however, a couple of distinctions can be made for 'in' and 'on'.

Physically, 'in' is used for indicating something is contained within something else.

*The description was **in the file** given to the network manager.*

While 'on' is used for something on a surface or just above.

*The devices were left **on the table** for the students to inspect.*

Also, use 'in' for months and years.

*This occurred **in** June.   This occurred **in** June 2014.   This occurred **in** 2009.*

And use 'on' for specific dates and days of the week.

*This occurred **on** June 15th 2014.   This occurred (~~in~~) **on** Tuesday.*

When referring to position in a diagram or on a screen use the following:

> ***at** the top   **in** the middle   **at** the bottom*

Prepositional phrases

Given their rather abstract nature and the numerous definitions they attract, the most effective way to learn prepositions is to become familiar with the words and phrases they form relationships with. Those terms most commonly found in engineering papers and most often misused are listed here by the preposition required:

at (full/low/high) power – *This is assuming the converter was operating in (**at**) full power prior to this.*

at

at (low/high/maximum) frequency – *These can be used in (**at**) both high and low frequency.*

run/travel at (speed/velocity) – *The wave spreads out and travels **at** a specific velocity…*

*They will not have access with (**access to**) any relevant information.*

to

> But when access is a verb it is not followed by 'to'.
>
> *The user accesses to the system in time to view this message.*

*This is the maximum number of nodes assigned in (**assigned to**) one cluster.*

*The user will then be able to connect with (**connect to**) the internet.*

*The optimization in (19) is equivalent with (**equivalent to**) the following:*

*This is identical with (**identical to**) the earlier fault.*

*We were successful in keeping this impact in a minimum (**to a minimum**).*

*This is in response with (**in response to**) the queries covered earlier.*

*…which works similar with (**similar to**) the photodiode.*

*Networks based on this are susceptible with (**susceptible to**) many threats.*

> *The longest part only reaches to 15 cm, so the sensor will not detect this.*

There is no delay ~~associated to~~ (**associated with**) *the current signals.*

with

Hopefully, this can be ~~combined to~~ (**combined with**) *efficient implementation.*

The anchors can ~~communicate to~~ (**communicate with**) *each other via the above mechanisms.*

Orion Ltd (Canada) ~~in collaboration of~~ (**in collaboration with**) *C____ University (China)…*

It was ~~equipped~~ (**equipped with**) *a 50,000 rpm electrospindle and…*

Clearly this ~~interferes on~~ (**interferes with**) *the signal.*

This demonstrates how the isocontour can ~~interact to~~ (**interact with**) *the square.*

In order to save time, it was ~~composed by~~ (**composed of**) *centroid nodes…*

of

It ~~consists with~~ (**consists of**) *temporary storage and more permanent storage.*

> A common error is no particle being used: *This* ~~consists~~ (**consists of**) *a single node and…*

We also acknowledge the ~~existence to~~ (**existence of**) *signal-dependent interference.*

This served as key referral material ~~in support to~~ (**in support of**) *my dissertation.*

The development looks to **take advantage** ~~from~~ (**of**) *the new SOAs.*

…largely ~~aided with~~ (**aided by**) *changing the direction of the flow.*

by

The reinforcement costs in the networks are ~~driven with~~ (**driven by**) *demand.*

Fig. 6. shows the new network ~~generated on~~ (**generated by**) *the centroid nodes.*

This is ~~indicated with~~ (**indicated by**) *a dashed line in the model.*

These have been ~~supplied from~~ (**supplied by**) various local manufacturers.

The CBs here are ~~triggered from~~ (**triggered by**) the busbar protection.

Here, the two aggressors will likely ~~compete on~~ (**compete for**) dominance…

> **for**

The ~~demand of~~ (**demand for**) power in this area will only increase in the future.

This traffic is ~~destined at~~ (**destined for**) a different subnet.

These limits demonstrate the ~~need of~~ (**need for**) alternative measures.

> These components represent ~~for~~ a practical grid system.

They are ~~responsible of~~ (**responsible for**) system security and routine maintenance.

Our method is ~~suitable to~~ (**suitable for**) the latter frequency channel.

The conditions have a great ~~effect in~~ (**effect on**) the electricity demand.

> **on**

These have yet to be ~~installed to~~ (**installed on**) the system.

Once uploaded, this will be visible ~~in the screen~~ (**on the screen**).

This occurs ~~in~~ (**on**) some occasions and usually when the system is idle.

They have displayed this information ~~in~~ (**on**) their website.

This is normally heard ~~at~~ (**in**) the background.

> **in**

One way is to extend the cluster ~~on both directions~~ (**in both directions**) as shown in Fig. 5.

This has been a key breakthrough that is now widely used ~~on the industry~~ (**in the industry**).

~~On normal mode~~ (**In normal mode**) these would be automatically loaded.

It probably indicates an unsolicited report has been sent ~~with~~ (**in**) **response to** a query.

These four phases will need to be implemented ~~at sequence~~ (**in sequence**).

CHAPTER 11

Clauses

Introduction

Every sentence has a main clause that contains a subject and a verb. Main clauses are also known as independent clauses because they can stand on their own without needing any additional information. They represent a complete idea or thought.

We tested the accuracy of our classifier.

A dependent clause depends on the independent (main) clause for its meaning so it cannot be used on its own. It is only part of a sentence and not a complete thought.

When the classifier was tested

Types of clause

There are two kinds of dependent clause: conditional and relative.

—conditional clauses

Conditional clauses talk about things that will, could or might happen now or in the future. They can also talk about things that could have happened in the past. They usually begin with 'if' or 'unless'.

If we begin now   *we might be able to transfer most of the data by the end of the day.*

It will not be able to function properly   **unless all nodes are available.**

—relative clauses

Relative clauses start with relative pronouns such as *that, which, whichever, who, whoever,* and *whose.*

They looked at the traffic   **that was building in the first network.**

There are two types of relative clause:

- A restrictive or essential relative clause influences the meaning of the sentence and cannot be left out.

  He was the researcher **who developed this model.**

- A non-restrictive or non-essential relative clause just gives extra information to the reader and if it is left out the sentence will still make sense.

  There were seven people working on the project, **which was an appropriate number.**

A comma always separates a non-restrictive relative clause from the main clause. If the relative clause is in the middle of the sentence then commas can be used on either side.

We turned on the device, **which took about ten seconds**, *and began the test.*

Take care

One of the most common mistakes occurs with the conjunction 'although'.

Many writers put a comma after 'although' when it is part of a dependent clause at the start of a sentence.

▸ *Although, our approach is similar, the requirements for live streaming applications differ.*

If a comma is used directly after 'although', it is difficult for the reader to follow the meaning of the sentence. A comma indicates a pause and there should not be one here.

'although' can be used in a non-essential relative clause between commas like this:

*The literature on this topic, **although** extensive, is widely scattered.*

Do not use 'but' to link to the main clause when beginning with 'although' (or 'even though').

Although the sample was large, ~~but~~ the results proved inconclusive.

Conjunctions

There are two types of conjunction used with clauses: coordinating and subordinating.

Coordinating conjunctions join independent clauses to form a single sentence.

and, but, for, nor, or, so , yet

This algorithm behaves perfectly in reconstructing the input signal, but the pitch shift fails to appear in this signal.

Subordinating conjunctions are used to begin dependent clauses.

after, although, as, because, even if, even though, if, since, unless, whereas...

A message at this stage is not possible, unless a new system is developed.

Conditional clauses

Dependent clauses that begin with 'if' or 'unless' are known as conditional clauses. This is because a certain condition must be met before the action in the independent or main clause can occur. These conditions can be probable, possible in theory or even impossible.

> *Conditional sentences may involve a prediction or opinion, or a clear intention to do something depending on a particular situation or action.*

Probable – *If they move quickly, they will take advantage of these favourable conditions.*

A clear intention now or in the future

Possible – *If they moved quickly, they would take advantage of these favourable conditions.*

An imaginary situation now or in the future

Impossible – *If they had moved quickly, they would have taken advantage of these favourable conditions.*

A remembered or imagined situation in the past

The main error with 'probable' conditionals is using 'will' in the if-clause instead of the main clause.

If they will go to the network manager, they will receive more detailed instructions.

Comparative construction

A rather curiously formed but nevertheless frequently applied construction is this comparative one:

The greater the usage, the higher the charges.

Both parts begin with 'the' and the verb 'to be' is often omitted. These are the four most common misconstructions:

The bigger is the system the less effective is the security.   ✗

The bigger the system is, the less effective is the security.   ✗

The bigger the system, less effective the security.   ✗

The bigger system, the less effective the security.   ✗

The bigger the system, the less effective the security.   ✓

Q and A

Q – When should I use 'that' in a sentence and when can I omit it?

A – It is always difficult to know when to leave out 'that' and when to retain it in a sentence.

When the 'that' is attached to the object of a noun clause it can be omitted.

*The users entered the passwords **that** they had been given.*

Read the sentence with and without 'that'. This should determine whether it can be omitted or not.

When 'that' is acting as the subject in an adjective clause (meaning it is part of a phrase modifying the noun) then it cannot be omitted.

*The next step is to reduce the number of channels (**that**) should be scanned.*

Only...

There are two constructions involving 'only' that can pose problems. The first one is a conditional that begins 'only if'.

When 'only if' begins a sentence the subject and verb are inverted (the same applies to 'only then' at the end of a sentence).

Only if the marginal costs are higher than the fixed costs they will make a purchase.   ✗

*Only if the marginal costs are higher than the fixed costs **will they** make a purchase.*   ✓

The second is when a sentence begins with 'not only'. Again, note the difference between the two errors and the correct format. The first part is inverted but the second part is not:

Not only ~~they must~~ have knowledge of the keys, but ~~also they must~~ know when to use them.

Not only must they have knowledge of the keys, ~~they also know~~ when to use them.

*Not only **must they** have knowledge of the keys, but **they must also** know when to use them.*

Another construction is 'not only….but also'.

*This is liable to depend **not only** on its value **but also** its history prior to time t.*

Example errors

Based on the aim of creating a bi-directional link between the two. It is necessary to analyse the links found in the previous…

Because the project is only interested in the data of the radar system, so the function of the camera control will not be considered

Here we have a dependent clause standing on its own. It is not a complete sentence, only an introductory phrase, so a comma is required not a period.

A coordinating conjunction is not necessary when linking a dependent clause with an independent one.

Do not use a semicolon to separate a dependent clause from an independent one.

Whereas the other statement indicates encrypted traffic; this statement specifies unencrypted traffic.

The delay spread can be small and then the frequency channel is flat with little difference between weights.

The two independent clauses do not coordinate well. The first clause should be changed to a dependent one to act subordinately to the second part.

When the delay spread is small, the frequency channel is flat with little difference between weights.

Though these networks can be distinguished from other kinds of ad hoc networks. Due to the nature of…

Again, here we have a dependent clause standing alone. It should be attached to the clause that came before it.

Text mining is a field of data mining, which offers potential for dynamic and rich analysis.

> Here a clause has been broken up too early with a comma and 'which'. This implies a non-essential clause when in fact the main clause should just continue with 'that' and no comma.
>
> *Text mining is a field of data mining that offers potential for dynamic and rich analysis.*

Because there are often several different approaches for handling the same situation, however it is necessary to have an elaborate network management system.

> Here, the use of both 'because' and 'however' creates confusion in the clauses. In this particular context 'however' is an unnecessary link going into the main clause and should be removed.

Since the data sampling rate of GMS-06 is quite high, ~~but~~ the frequency of 50 Hz power frequency is relatively low.

> Again, no need for the linking word here.

Chapter 12

Prefixes

Introduction

A prefix is a tag found at the beginning of a word. It serves to modify the meaning of the word. Although they have specific meanings it is not always possible to judge what particular prefix should be used and a dictionary should be consulted if in doubt. The definitions of some common prefixes are provided here for reference:

de–	taking something away, opposite	**dis–**	reverse, opposite
in–	not, negative	**mis–**	bad, wrong
non–	not, absence of something	**re–**	again, repeatedly
un–	not, opposite (not always a negative meaning)		

See Chapter 22 for a list of unit prefixes.

Forms and errors

Errors often occur when using the prefixes listed above as these sets of letters commonly form new words from existing ones (unlike say the prefix 'para' in 'parallel' whose root is not a word) and writers may come unstuck guessing or confusing the form. These examples should cover most of the key terms in the field:

de–   (deconstruct, decompress, decouple, decrypt, decouple, denoise)

decode   *It would take too long to ~~uncode~~ (**decode**) this series.*

dis–   (disable, disadvantage, discharge, disconnected, discontinuous, displacement, dissimilar, distrustful)

disassociation   *The data utility can be achieved by this ~~unassociation~~ (**disassociation**).*

See A–Z for contiguous

im–   (impedance, impossible, impractical, imprecise, improper)

imbalance   *This will help to prevent any system ~~unbalance~~ (**imbalance**).*

However, 'unbalanced' is an accepted adjective.

in–   (inaccessible, inactive, independent, inefficient, inelastic, inexpensive, inflexible, invulnerable)

inaccurate   *When it is not linear, the range estimation will be ~~in accurate~~ (**inaccurate**).*

incompatible   *We would then check if there was ~~uncompatible~~ (**incompatible**) software.*

input   *The ~~imput~~ (**input**) voltage is dependent on the array design.*

invalid   *The next stage is to identify any ~~in valid~~ (**invalid**) routes.*

See A–Z for insecure/unsecured

non–   (noncausal, nonnegative, nonparametric, nonuniform, nonzero)

nonlinear   *It has the capability to handle dislinear (**nonlinear**) relations between variables.*

nonrandom   *These exist as unrandom (**nonrandom**) binary codes.*

See A–Z for contiguous

re–   (reactive, reallocate, rearrange, reassemble, reassess, reassign, recall, reconfigure, reconnect, reconstruct, recurrent, redirect, retransmit, retune, rewrite)

reboot   *If this happens we need to boot the system again (**reboot** the system).*

un–   (unaligned, unavoidable, unbounded, unclear, uncontrolled, uncorrelated, undesirable, undetectable, undesirable, undirected, undistorted, unexpected, unexplained, unfiltered, unknown, uninitialized, unintended, unmanned, unmasked, unmodulated, unpredictable, unproven, unreached, unreadable, unreliable, unrestricted, unroutable, unsuitable, unsolicited, unsupervized, untimed, unused)

Also: uncertainly

uncertainty   *Incertainty (**Uncertainty**) exists with long term forecasting…*

uncontrollable   *The converter becomes an incontrollable (**uncontrollable**) diode bridge…*

unfeasible   *It would have been infeasible (**unfeasible**) with older networks.*

unlike   *This is groundwave propagation dislike (**unlike**) other types of radar systems.*

unstable   *In the MEEP simulation it will become instable (**unstable**).*

'dislike' is a verb

See A–Z for contiguous   See A–Z for insecure/unsecured

If there is a valid and commonly used prefix (check that the prefix can be attached to that particular word if unsure) for the opposite, alternative, negative, or the repetitive meaning then use it. It is always preferable to adding 'not' 'or 'again'.

*The relay receives this version of the sequences and then ~~transmits again~~ (**retransmits**) this noisy version.*

*They would expect to have ~~not restricted~~ (**unrestricted**) access at this time.*

Not knowing a prefix is even more apparent when the sentence appears to be positive, producing a cumbersome and awkward reading.

This makes it not a suitable method.   ✗

This makes it an unsuitable method.   ✓

Sometimes a choice can be made between a negative beginning to the sentence or a prefix indicating negation.

All other protocols are not affected   ✗

No other protocols are affected   ✓

All other protocols are unaffected   ✓

Hyphens after prefixes

It is general practice to refrain from using a hyphen after a prefix.

That said, hyphens will be required for the following:

all-   self-   ex-   half-   quarter-

*One advantage is that it is **self-regulating**.*

*The beamwidth should be able to cover this **half-space** area.*

*We will refer to this as an '**all-sided** relationship'.*

*An **all-encompassing** system is unrealistic at the current time.*

And if the word has two different meanings, a hyphen can be used to differentiate.

recover – *The main objective would be to **recover** the file.*

re-cover – *This can be used to **re-cover** the components if the material has been worn away.*

Section II

Style and punctuation

CHAPTER 13

Style: Clarity and brevity

Introduction

When writing an engineering paper the focus should be on communicating your ideas in the most effective way possible. Although identifying unnecessary phrases and eliminating wordiness in your work can be a challenge, there are some terms you should reject immediately and others that can be efficiently modified.

Redundant and unnecessary terms

Improve the quality of your writing by removing unnecessary words and using shorter phrases:

~~still~~ continues	**continues**	*Even now this ~~still~~ persists/ continues.*
few ~~in number~~	**few**	*There were also undecided people but they were few ~~in number~~.*
~~new~~ innovation	**innovation**	*It is ~~a new~~ (an) innovation that has seen many...*
may ~~possibly~~	**may**	*This may ~~possibly~~ lead to further power losses.*
~~brand~~ new	**new**	*The company released a ~~brand~~ new product.*
~~continue to~~ remain	**remain**	*This should ~~continue to~~ remain the situation throughout the testing phase.*
return ~~back~~	**return**	*It meant a return ~~back~~ to the old system.*
proceed ~~forward~~	**proceed**	*The subscriber proceeds ~~forward~~ at this point.*
~~very~~ critical	**critical**	These are strong adjectives that require no emphasis.
~~very~~ crucial	**crucial**	
~~very~~ necessary	**necessary**	See also Chapter **9** excessive adjectives.
consider ~~about~~	**consider**	discuss ~~about~~ **discuss**
increases ~~up~~	**increases**	reduces ~~down~~ **reduces**

—long-winded phrases

Some long phrases contain barely any information and can be simplified, sometimes down to a single word.

another number of	**additional**	as it has been shown above **as shown above**

at the present time **currently**	by means of which **whereby**
call back to mind **recall**	do not provide any contribution **do not contribute**
due to the fact that **because**	
in a way that **whereby**	during the course of **during**
in the event that **if**	in the case of **for**
let us not lose sight of the fact **remember**	is capable of/is able to **can**
	with the exception of **except**

Q and A

Q – is it 'join together' or just 'join'?

A – 'together' can often be excluded as these examples demonstrate:

When the attackers unite ~~together~~ they are difficult to block.

This meant the two could now collaborate ~~together~~

...and then they will be able to co-operate ~~together~~

They were then joined ~~together~~ to create a single standard.

...and these will link ~~together~~ to form a stronger connection.

The two models can be combined ~~together~~.

In this case, classes i and j do not overlap ~~together~~ in the feature space.

—acronyms and abbreviations

When using acronyms and abbreviations, first define them in full and then refer to them each time in the abbreviated form.

It is faster than the more common dynamic random access memory and has greater reliability. Dynamic random access memory (DRAM) can support access times of approximately 10 ns while...   ✗

It is faster than the more common dynamic random access memory (DRAM) and has greater reliability. DRAM can support access times of approximately 10 ns while...   ✓

Although acronyms defined in the nomenclature (list of terms at the beginning of a paper) perhaps do not need to be defined again in the main text, a definition is helpful for the reader in the first instance. Define the acronym the first time it is used in both the abstract and the main body. The following acronyms and abbreviations do not need to be defined in the main text as they are considered standard knowledge:

Example: ac *However, the ac (~~alternating current~~) resistance changes significantly when the frequency increases.*

A/D	FM	IR	MHD	PC	ROM
AM	FTP	*I-V*	MIS	p-i-n	RV
ATM	GUI	JFET	MMF	p-n-p	SIR
B—E	HF	JPEG	MOS	PML	TE
CD-ROM	HTML	LAN	MOSFET	PTM	TM
CPU	HV	*LC*	MOST	RAM	UHF
CRT	IF	LED	MPEG	*RC*	UV
CV	IGFET	LMS	n-p-n	RF	*V-I*
dc	IM	*LR*	OD	*RL*	WA
FET	I/O	MESFET	OOP	rms	

—repetition of term

It is important to shorten terms and utilise pronouns when an abbreviation is not available. This seems obvious enough but it is surprising how much repetition occurs in writing.

An analytical linear model *was developed initially. ~~The analytical linear model~~* **(The model)** *was verified by comparing it with the developed model.*

The conventional 6-transistors (6T) SRAM cell *is widely used in digital systems. ~~The conventional 6-transistors (6T) SRAM cell~~* **(It)** *uses six transistors to store and access…*

Using this tool helps analysts decide which routes ~~analysts~~ **(they)** *should take.*

And when an abbreviation is available and has already been introduced, there is no need to keep reminding the reader.

The DSA ~~which stands for digital signature algorithm~~ has also meant that…

The DSA ~~for digital signature algorithm~~ should be able to…

Position of the subject

It is always good practice to place the most important point at the beginning of the sentence. Note the position of the subject here:

To construct a confidence interval for the normal distribution in small samples, we use likelihood-based approaches.   ✗

We use likelihood-based approaches to construct a confidence interval for the normal distribution in small samples.   ✓

Of

The 'of' phrase can needlessly lengthen a sentence and is overemployed by some writers; nevertheless, it is appropriate for indicating possession when the phrase is referring to an inanimate subject and also for adding clarity to a long noun phrase.

The distributed antenna system's quality...

The quality of the distributed antenna system...

Avoid the habit of breaking up compound nouns unnecessarily:

✗	✓
where the value of the tag is...	*where the **tag value** is...*
This would involve reassessing the policy of routing...	*...the **routing policy**.*
Another factor is usage of system	*Another factor is **system usage**.*
They also studied detection of faults	*They also studied **fault detection**.*

But sometimes it is a recognised phrase and should not be turned into a compound.

Denial of service *can also affect the network indefinitely.*

When forming a plural 'of' phrase, only the first word is pluralised.

level of intensities	*marker of times*	*school of thoughts* ✗
levels *of intensity*	***markers*** *of time*	***schools*** *of thought* ✓

Conjunctions

Conjunctive adverbs (furthermore, however, meanwhile...) should be used sparingly. Compare the systematic use of adverbs as clause starters in the first passage with their prudent absence in the second.

A typical FMCW radar model is depicted in Fig. 2.1. Ultimately, it transmits a high-frequency signal with a continuously changing carrier frequency over a chosen frequency range on a repetitive basis. Moreover, this is advanced by means of a voltage-controlled oscillator. On the one hand, the frequency difference is obtained by a mixing process; on the other hand, the frequencies of the received echoes are recovered by spectral analysis of the mixer output.

A typical FMCW radar model is depicted in Fig. 2.1. It transmits a high-frequency signal with a continuously changing carrier frequency over a chosen frequency range on a repetitive basis. This is advanced by means of a voltage-controlled oscillator. The frequency difference is obtained by a mixing process, and the frequencies of the received echoes are recovered by spectral analysis of the mixer output.

Using lists

Lists are an effective way to condense parallel information. They serve to emphasize the information and display it clearly for the reader. Those that are embedded in the text should be in the following format:

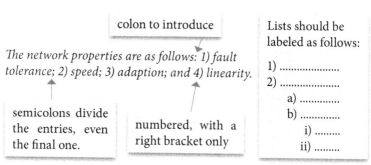

colon to introduce

The network properties are as follows: 1) fault tolerance; 2) speed; 3) adaption; and 4) linearity.

semicolons divide the entries, even the final one.

numbered, with a right bracket only

Lists should be labeled as follows:

1) ....
2) ....
  a) ....
  b) ....
    i) ....
    ii) ....

Lists that are displayed take this format:

The reasons for designing these networks:

1) They provide a sufficiently low impedance path.

full stop after each entry and capital letter to start

2) They retain system voltages within reasonable limits during fault conditions.

3) They ensure that step and touch potential voltages are kept to a minimum.

Lists can be made up of incomplete sentences. In this case the entry does not take a capital letter and is separated by a semicolon.

1) to analyse the problem;

2) to identify an algorithm;...

Lower case

Try to keep the entries to a parallel structure (all in –ing form, infinitive form...)

1) to analyse the problem;

2) identifying an algorithm;

Nomenclature lists (lists before the introduction that define symbols) are written,

f1 = supply frequency

f2   the slip frequency

p   is the pole pairs of the stator windings

s   motor per-unit slip

s(t)   wavelet transform of function

u   shift

$\phi(t)$ mother wavelet

Do not use a dash or equals sign to introduce the definition. Do not use articles or 'is' in the explanation.

When ending a summary do not use 'at last' – it sounds as though you are relieved it is over.

*...third, we will address the main concerns of the network users; at last, (**last,**) we will offer a conclusion based on these discussions.*

You can also use:

finally   lastly (with *firstly, secondly...*)

Do not introduce a list and then add i.e. or e.g.

This is divided into four stages i.e.

CHAPTER 14

Style: Voice and verb choice

Introduction

Technical writing has long been associated with a method/procedure-driven narrative and the passive voice. Increasingly though, writers are preferring to – and being instructed to – write in an active voice for presenting and describing their research. This brief guide will illustrate the structural differences and the areas of importance.

There are two voices in English, the active voice and the passive voice.

The subject performs the action in a sentence with an active voice:

The engineer plays an important role in this process.

But in certain situations the word order can be altered and the sentence can be changed from 'active' to 'passive'. When this happens the subject goes from performing the action to being acted upon by the verb.

active	*The engineer also increased safety.*
passive	*Safety was also increased by the engineer.*

Pronouns are removed when we change from active to passive:

We note an error here.

An error was noted here.

I will use a pie chart.

A pie chart will be used.

Even if the active sentence is in the present tense the passive sentence will still use the past participle.

For the passive form the following things happen:

- The subject moves to the end.
- 'Safety' has now become the subject of the sentence.
- The preposition 'by' now follows the past participle ('*increased*').
- A form of the verb 'to be' ('*was*') comes before the past participle.

Checklist for changing active to passive:

- Move the subject of the active sentence to become the object of the passive
- Use the correct forms of the verb 'to be' before the verb in the passive sentence
- Change the verb to the past participle form
- Choose whether to include the object at the end of the passive sentence
- If so, put 'by' before the object

Tense

When changing a sentence from active to passive the tense does not change. If the active sentence was in the past tense use this for the passive sentence as well:

The designer changed the concept.

The concept was changed (by the designer).

Because the –ed form is generally associated with the past tense, errors can occur when the passive is used for a present action.

*The output of the laser diode is ~~record~~ (**recorded**) in the amplitude modulator...*

—the verb 'to be'

Note which form of the verb 'to be' to use for the passive voice in the progressive and perfect tenses:

progressive tense: (add 'being')	perfect tense: (add 'been')
The designer is changing the concept.	*The designer had changed the concept…*
The concept is being changed (by the designer).	*The concept had been changed…*

Take care

Using the passive voice can sometimes lead to a subject that is too complex and too far away from the verb.

__A solution__ for reducing operational costs and ensuring the efficient and economic utilization of power generation __has been found__. ✗

__A solution has been found__ for reducing operational costs and ensuring the efficient and economic utilization of power generation. ✓

Merits of both voices

The **active** voice is preferred by most writers because the sentences are more appealing for the reader and the writer can get a point across in a direct and clear way. Another effect is that the subject takes responsibility for the action.

They changed the system too quickly.

Passive sentences are useful for writers who wish to remain neutral. They are traditionally employed when the subject is less important than the process being described. Stylistically they can provide variety and an alternative to beginning every sentence with a personal pronoun.

> *We used this as the basis for our model as suggested in [11]. The model __was designed__ to reduce feedback time and…*
>
> Rather than '*We modelled the design…*'

—restrictions on the passive voice

Only sentences with a direct object can change voices. In other words, only verbs that are transitive (those that take objects) can be reworked into the passive form.

We made an error on the third electrical circuit.   active

An error was made on the third electrical circuit.   passive

> Note: *The message arrived at 6pm.   At 6pm the message arrived.*
>
> This is not an example of active and passive. The prepositional phrase has just been moved to the beginning of the sentence.

Sentences with the auxiliary verb 'to have' as the main verb cannot be transformed into passive either.

The administrators had a different role.   ~~A different role was had by the administrators.~~

Nominalization

Nominalization occurs when a verb (or adjective) is replaced by the noun form.

investigate – verb   investigation – noun.

We investigated *some critical parameters to improve the static noise margin (SNM).*

An investigation was carried out of *some critical parameters to improve the static noise margin (SNM).*

The first sentence is concise and takes only two words (*we investigated*) to inform the reader that there was an investigation. The second sentence is wordy, vague and takes six words (*an investigation was carried out of*) to convey this information.

We made an evaluation of these three storage methods.   ✗

We evaluated these three storage methods.   ✓

In [15] they took measurements of each component and…   ✗

In [15] they measured each component and…   ✓

Our results were found to be in agreement with [9] and [13]. ✗

Our results agreed with [9] and [13]. ✓

Nominalization produces verbals that add very little to a sentence. By omitting the unnecessary terms, the sentence can be almost halved in size.

The first step will be to conduct an evaluation of the whole sub-threshold design.

Instead of using 'there is/are' or 'there was/were', simply bring the subject back alongside a verb:

There were six problems in relation to the initial scheme.

The initial scheme had six problems.

First, we will evaluate the whole sub-threshold design.

But using nouns in this way should not be avoided altogether. They can add variety to the sentences and prevent repetition, and can also provide a link to a previous idea or action.

We discussed the optimum size of the network with the network managers. A discussion of the key security features also took place to ensure...

Verb strength

Once a verb has been chosen it can take on different degrees of strength depending on whether it is used actively, passively, as a verbal or nominalized. As a rule the stronger the verb form, the clearer and more dynamic the sentence.

If we extract whole images...   STRONG

If whole images are extracted...

With extracted whole images...

With extraction of whole images...   WEAK

Verbs can be weakened even more by hesitant phrases (in bold).

This will allow the user to communicate with our system.

*This **seems to** allow the user to communicate with our system.*

*This **might** allow the user to communicate with our system.*

CHAPTER 15

Tense

Introduction

Tenses are used to mark the time of an action or an event.

Initially, we can split the tenses into present, past and future: (verb – to change)

present: *He changes*　　past: *He changed*　　future: *He will change*

In the future tense the verb form stays the same. We use 'will' along with the normal or dictionary form of the verb. The three examples above are all in the simple aspect. There are other aspects that assist us in being more exact about the time of an event/action.

perfect   progressive

These aspects are used with the three tenses as follows:

present perfect:
action that begins in the past but continues into the present or the effect continues.

past perfect:
action in the past and completed before another action.

future perfect:
action that will have been completed at a specific time in the future.

present progressive:
action happening at the moment or fixed in the near future.

past progressive:
action that was happening at some point in the past.

future progressive:
action that will be happening at some point in the future.

present perfect: *It has changed*   present progressive: *It is changing*
past perfect: *It had changed*   past progressive: *It was changing*
future perfect: *It will have changed*   future progressive: *It will be changing*

There is also a perfect progressive form:

present perfect progressive: *It has been changing*
past perfect progressive: *It had been changing*
future perfect progressive: *It will have been changing*

Take care

The past perfect can be overused and in fact is quite limited. It is only used to refer to an event in the past if you are also mentioning another event in the past. When describing one event that took place in the past just use the past tense.

We changed the arrangement to
improve connectivity. ⟵ one event (past)

When the users logged on the
operator had changed the settings
to restrict their access. ⟵ two events (past perfect)

Changing tenses

As a general rule tenses should remain consistent in a sentence or paragraph unless the time of the action being described or the viewpoint of the author/researcher changes.

*We **obtained** an SDP in the fixed transmit sequence case and **devised** new algorithms to synthesize the transmit sequences.*

But there are many occasions where the tense can be changed in a sentence, especially when the main verb is in the present or future tense. If the main verb is in these tenses, the subordinate verbs can be in any tense.

*Our findings **prove** that the ratio **will remain** constant even after the constraints are added.*

An example of a subtle shift in tense is when the sentence begins with a time expression (such as *when, before, after, if, unless*); the first part of the clause is in the present tense and the second in the future.

| present | future |

When they search the database   recommendations will appear.
When they ~~will~~ search the database…

Confusion arises when the writer attempts to follow guidelines asking for, say, the literature review to be written exclusively in the past tense. The temptation is then to write every verb in the past tense.

*The analysed literature **varied** in scope and breadth... in general terms, a radar system ~~used~~ (**uses**) electromagnetic waves to identify the range and direction of objects.*

Using the past tense for the literature review means that reporting verbs and any particular experiments or research are explained in the past tense. In the example above a well-known fact is stated (a radar system uses electromagnetic waves...), so the verb must be in the present tense. If something is still true today or is a current event then the past tense is inappropriate. So even if a particular tense is to be favoured, each verb instance should still be assessed for appropriate usage.

Verbals can take a different tense to the main verb in the past and perfect forms. Here the action takes place before that of the main verb:

*Low voltage circuits **working** well in the previous experiment **supported** our argument.*

Tense selection

In academic writing, when reporting on events generally or how something was done the past tense is usually used.

*As previously reported, [12] and [14] **conducted** an experiment to test this theory.*

The present tense is useful to show that the event or finding is relevant and accepted, because it can project a sense of significance or importance.

*Interestingly, [14] **reveals** that it **is** because these components have a smaller influence.*

Descriptions in the illustrated analysis (the results) of your own tests are usually in the present tense.

*From our analysis, the values of SNM **decrease** as the supply voltages **drop**.*

When summarising or concluding, the past tense is the clear choice. A common mistake is to use the present tense when referring to earlier sections of a paper.

The previous chapter ~~looks~~ (**looked**) *at the problems in integrated systems and then* ~~assesses~~ (**assessed**)...

When referring to tables and diagrams that follow, it is logical to use the present tense and not the past.

Future events

One ambiguity is that we do not always have to use the future tense to discuss future plans or events.

We know that 'will' is used before the verb to discuss future events generally.

The team **will decide** *whether to continue with the design.*

But when we know about the future we can actually use the present tense. Here something has been arranged and the present tense is used.

They **have** *another meeting next week.*

Future plans can also be mentioned in the present progressive.

They **are planning** *to add another layer to the middleware.*

A key point to remember when using the progressive tense is that only certain types of verbs can be used. These verbs are known as dynamic verbs and they relate to activities that can begin and finish.

> **Examples of dynamic verbs**   ask, call, change, feel, help, learn, listen, play, read, work, write

Chapter 16

Time and duration

Introduction

This section outlines the key prepositions and adjectives that are used and misused when expressing time. It also covers the tense in which useful time phrases are written and offers guidance on the always awkward sentence openers – which invariably contain a time expression.

Opening statements

The opening sentences of many papers include some kind of time reference, usually about how something has grown or changed over the years, when something was created or implemented or that research has increased recently in a particular area. The following issues should be recognised:

When discussing a recent event or trend,

~~In recently~~   **Recently,**

~~In the recent times/years~~   **In recent times/years,**

~~In nowadays~~ ~~Nowadays~~   **Nowadays,** ◄

Use present perfect or present progressive tense for the first two terms and simple present for the third.

Do not use 'nowadays' in a possessive way. It cannot be used like 'today's'.

*This is more appropriate for ~~nowaday's~~ (**today's**) internet.*

Nowadays, it is common to see this type of system.

*In recent years the demand ~~increases~~ (**has increased**) dramatically.*

Use past or last instead of recent or latest when referring to the previous ten years (decade).

*There has been more progress made this year than in the ~~latest~~ (**past**) decade.*

*During the **last** decade, research on image retrieval (**has**) evolved.*

Use 'recent years' but not 'recent decades'. Given the fast-changing nature of industry and research there is nothing 'recent' about twenty or thirty years ago. Use 'past few' for decades and note the verb form.

*There ~~was~~ (**has been**) an increase in capability in the **past few decades** and this has resulted in...*

*As a key component of many economic activities, the demand for radio spectrum ~~is~~ (**has been**) rising **in recent years**.*

Time expressions
—tense (example verb: to access)

SIMPLE FUTURE

We will access this data **next**

We will access this data **next time**

We will access this data **in a week**

We will access this data **tomorrow**

PRESENT PERFECT PROGRESSIVE

We have been accessing... **all morning**

We have been accessing... **since 3pm.**

We have been accessing... **for two days**

We have been accessing... **for the last four hours**

SIMPLE PRESENT

...access the internet **every day**

...**always** access the internet

...**sometimes** access the internet

...access the internet **once a week**

...**never** access the internet

PRESENT PROGRESSIVE

They are accessing... **now**

They are **currently** accessing...

They are **temporarily** accessing...

They are accessing... **this week**

SIMPLE PAST

...accessed the network **yesterday**

...accessed the network **last week**

...accessed the network **earlier today**

...accessed the network **two days ago**

PRESENT PERFECT

They have accessed... **several times**

They have accessed... **since last week**

They have accessed... **recently**

They have accessed... **in the last month**

—since/from

'*from*' can indicate a specific place or time as a starting point.

***From** 2018, this will be mandatory for all devices.*

Use '*from*' to also indicate the first of two specific points (with '*to*' or '*until*').

*...**from** the signal to the base station.   ...**from** six until seven.*

As a preposition 'since' is used to mean continuously from a certain time.

*This has been occurring **since** the parameters were changed.*

The use of '*since*' and '*from*' can therefore be contrasted by the time of the event.

*We have been studying this **since** April.*   present

*We will be analysing this data **from** October.*   future

Remember, for time-related sentences do not use 'since' with the simple present tense.

*Since 2005, version 5 ~~evolves~~ (**has evolved**) to include a source list.*

It can be used with different tenses as a conjunction meaning 'as' or 'given'.

Since it is affecting consumption, we take the necessary steps to...

—until/by

Use 'until' when the activity continues up to a specific end point.

*We will continue **until** the maximum transmission rate is reached.*

Use 'by' to set a time limit for an activity or situation.

*The web traffic will have peaked **by** 18:00.*

*The scheme should be finished ~~until~~ (**by**) the end of the month.*

—prepositions

Note the prepositions used in these time sequences:

For this session we measured the number of users

at *the beginning*   **in** *the middle*
at *the end*

'in the end' usually means after all or finally and describes an outcome or how something was eventually done. It is not a strict time expression.

We measured the linear loads of the system

at noon   at night

in the morning   in the evening   in the afternoon

And note these time phrases:
in time – eventually
over time – over the course of time
on time – punctual; at the specified time
at the same time   at a point in time

—adjectives

When using units of time as adjectives, make sure that the plural form is dropped and the singular is used with a hyphen.

This is scheduled to be a ~~thirty-minutes~~ **(thirty-minute) task**.

daily	*This will be carried out* ~~each~~ *daily*
weekly	
monthly	*I used* ~~month~~ **(monthly)** *data.*
quarterly	
yearly	*They also looked at* ~~year~~ **(yearly)** *projections.*

A ~~three months break~~ *affected the scheme.*   *A **three-month break**…*

Q and A

Q – What adjectives can I use with 'time' to describe the duration of something?

A – The following examples show which adjectives to use and which not to:

These operations will take a long time.

These operations will take a short time.

These operations will take a ~~small time~~. (small amount of time is Ok but wordy)

These operations will take a ~~big time~~.

CHAPTER 17

Titles

Introduction

This section presents the rules and regulations for producing a title for your work. As the first piece of text the reader sees, titles need to be relevant (reliable), effective (persuasive) and appropriately edited (professional).

Compare the following:

An overview of Fromal engineering in terms of Industrial software Development mainly Though the use of the SOFL Method

Formal Engineering for Industrial Software Development Using the SOFL Method

Title of the paper

The main title should inform the reader of the nature of the work as succinctly as possible. Only key terms should be used and adjectives and introductory statements avoided unless fundamental to the paper. Remember that the abstract will provide the necessary detail for the reader to fully assess the work.

Let's evaluate the two titles provided earlier:

needless introduction		inconsistent use of capital letters

A detailed overview of Fromal engineering in terms of Industrial software Development mainly Though the use of the SOFL Method ✗

two spelling mistakes	unnecessary padding	unnecessary noun phrase

consistent capitalisation

Formal Engineering for Industrial Software Development Using the SOFL Method ✓

ideal length	verb form instead of noun phrase

—capitalization

Every word in the main title should be capitalized except:

- short prepositions (such as *of, for, by, in, on, at*)

- articles (the, *a/an*)

- coordinating conjunctions (*and, or, for, but*)

- abbreviations of units (*mil, bit…*)

Longer prepositions are capitalized

A Study of Nonlinear Harmonic Interaction Between a Single Phase Line-Commutated Converter and a Power System

Only if an article is the first word of the title should it be capitalized

Titles of chapters and figures

Similar article rules apply for titles of chapters and figures.

Design issues affecting network

> Note how the articles are missing here where normally indefinite or definite articles would be used.

Fig. 3-5. Dual system with transmission cable

Q and A

Q – Is there a particular format to follow for headings?

A – Primary headings (section) should be centered and numbered using roman numerals not letters. Secondary headings should be numbered using letters and italicized. Headings after that should be numbered with a right bracket and then lower case letters used with only the first word of the heading in capitals.

IV RESULTS	1st heading
A Measurement of the Switch	2nd heading
1) Comparison:	3rd heading
a) Initial summary:	4th heading

CHAPTER 18

Spelling form

Introduction

The majority of international engineering journals and organizations use American spelling so this form is advisable, unless you are writing an internally-assessed paper for a British university. A comparative list of a few common terms is presented here for reference.

Differences between American English (AE) and British English (BE)

BE – italics **AE** – bold

—ize/ise ization/isation

For these verb and noun endings the AE form is –ize/ization while BE uses –ise/isation:

analyse	**analyze**	*authorise*	**authorize**
emphasise	**emphasize**	*generalise*	**generalize**
harmonise	**harmonize**	*initialise*	**initialize**
optimise	**optimize**	*organisation*	**organization**
paralyse	**paralyze**	*realise*	**realize**
specialise	**specialize**	*standardisation*	**standardization**
supervise	**supervize**	*utilise*	**utilize**

—l/ll

Verbs in AE contain one 'l' and for BE 'll'. For nouns it tends to be the opposite and relates to the first letter of the suffix. If it starts with a consonant then in AE the 'l' is doubled.

cancelled	**canceled**	*dialled*	**dialed**
enrolment	**enrollment**	*equalling*	**equaling**
instalment	**installment**	*modelling*	**modeling**
signalling	**signaling**	*travelled*	**traveled**

—iour/our

The 'u' is dropped from these terms in AE.

behaviour	**behavior**	*colour*	**color**	*favourite*	**favorite**
labour	**labor**	*honour*	**honor**	*neighbour*	**neighbor**

—log/logue

analogue	**analog**	*catalogue*	**catalog**	*dialogue*	**dialog**

—ce/se

AE makes no distinction between the verb forms of nouns ending in –ce or –se. In BE they are nouns and verbs respectively. Other nouns that end in –ce in BE may end in –se in AE.

defence	**defense**	*offence*	**offense**

practise (verb) **practice** (noun and verb)

license (verb) **license** (noun and verb)

—ter/tre

Certain words ending in –tre in BE are written –ter in AE

centre	**center**	*fibre*	**fiber**	*litre*	**liter**
lustre	**luster**	*Metre*	**meter**		

—also note the following:

acknowledgement	**acknowledgment**	*cypher*	**cipher**
fulfil	**fulfill**	*grey*	**gray**
judgement	**judgment**	*manoeuvre*	**maneuver**
routeing	**routing**	*skilful*	**skillful**
speciality	**specialty**	*vender*	**vendor**

CHAPTER 19

Capitalization

Introduction

Knowing when to use a capital for the initial letter of a word can be a guessing game for some, but there are straightforward rules. The use of capital letters in titles has already been discussed so here we are concentrating on the main text and the relationship with nouns.

General principles

Capitalization is considered a rare event in engineering papers. Capital letters should only be used for the following:

—proper nouns

Names of people and places will always take a capital letter.

*In [14], **Morgan** and colleagues discuss the performance improvement when using the combined technique.*

*This had been published in **Hyderabad, India** the year before.*

People's names that form laws, theorems and models will require a capital letter as well, but not usually the words that follow:

> Cayley-Hamilton theorem
> Coulomb's law
> Dennard scaling
> Gustafson's law
> Hopfield method
> Kalman filter
> Kirchhoff's laws
> Schur's inequality…

*The conductor size can be calculated using **Kelvin's law.***

*This was based on Armstrong ~~Oscillator~~ (**oscillator**) topology.*

*In [10] they simplified a naive **Bayes** classifier…*

And places that feature in these terms will also require capitals.

*One of the most common is **Monte Carlo** analysis…*

Names of computer languages and systems are capitalized.

Linux   Symbian   Windows Mobile

In fact, ***Perl** is another example of a scripting language.*

—titles and terms

The titles of IEEE publications should be entirely in small caps if referenced in the main text.

IEEE Spectrum/IEEE Transactions On Electron Devices

IEEE Transactions on Image Processing/IEEE Transactions on Power Electronics

The following are also entirely capitalized:

RC   RL   I-V   LC   S/N   (and italicized).

NOR   AND   OR   NAND   XOR

GO TO   DO   READ   WRITE   ON   OFF…

Note the capital part of these terms:

***O** ring*

***T** junction*

***Y**-connected circuit*

*class–**A** amplifier*

Take care

The inaccurate and seemingly arbitrary use of capital letters by some writers is ably reflected in the following passage. NONE of the terms in bold require a capital letter:

*Generators and **Inductors** are the two main elements in a **Power System** affecting the **Magnitude** of the **Fault Current**. They are the main **Elements** that will affect the unutilized capacity or fault current headroom of the circuit breaker. That is because these two elements are the most flexible ones as the demands increase day after day. The magnitude of the fault current partly depends on the internal impedance of the **Generators** and the equivalent impedance of the rest of the network; therefore, the **Fault** level will grow gradually when more and more DGs are connected to the **Distribution Network** (see 2.3 for an extended **Discussion** on this).*

CHAPTER 20

Colons, parentheses, dashes…

Introduction

Punctuation is used to organize writing, enhance readability and promote understanding. Too much punctuation or too little will make for an unpleasant reading experience, whereas the wrong punctuation will obscure and even completely change the meaning of a sentence.

We will require a radio, battery and hard, wired shifts.

> A comma is recommended for the penultimate item in a list.
>
> *…battery, and hard-wired shifts.*

We will require a radio battery, and hard-wired shifts.

The wrong punctuation can also make the sentence unreadable.

If it is not the opportunity to change frequencies, will be lost. ✗

If it is not, the opportunity to change frequencies will be lost. ✓

Colons and semicolons

Colons are two dots one above the other and are used to introduce things. They should only be used after complete sentences.

We will assess: their suitability and any negative effects. ✗

There are many variations such as: the number of tubes, the diameter of the… ✗

The colon can be used after 'as follows' or 'the following' to introduce a list.

The main results can be summarized as follows:

Semicolons are used to separate sets of independent clauses in complex sentences.

In this thesis only one system is studied for determining radio coverage; however, the study contrasts the proposed...

Semicolons are more powerful than commas and are effective for contrasting two clauses.

The first method produced sufficient simulation time and was able to determine unique values; the second method failed in both these tasks.

Horizontal lines

Hyphens and dashes may look the same but they perform very different functions.

- hyphen   – en dash   — em dash

-hyphens

Hyphens are the smallest of the three horizontal lines. They connect words and word fragments but in engineering papers should be adopted sparingly.

They are used,

in terms that are modifying a noun and coming before the noun

but not after the noun

*The Gaussian we applied was **bell shaped**.*

*We apply the **bell-shaped** Gaussian...*

if two words are modifying each other and then these words are modifying the final noun (or if the modifier is complex)

If the words include an adverb ending in –ly, a comparative/superlative or they are a pair of words not directly connected to the final noun, do not use a hyphen.

*This was a **newly installed system**.*

*Using **low-cost devices** would have its own pitfalls.*

*To improve the **source-to-end** delay...*

*Next we will discuss the **small, positive values**.*

if the prefix part of the word ends in a vowel and the next part begins with the same vowel

*The issue relates to whether we could **re-energise** it in sufficient time.*

> There are a couple of exceptions to this:
>
> *cooperate/cooperation*
> *coordinate/coordination*
>
> See Chapter '**12 Prefixes**' for further guidelines.

in compound verbs even when they are not being used as adjectives

*They could **water-proof** this for obvious safety reasons.*

when the modifier contains a present/past participle

*It also contains a **signal-limiting** property.*

> Phrasal verbs do not take a hyphen:
>
> It will **read out** after the user has taken the necessary steps.
>
> The noun forms are not hyphenated either.
>
> *We can then take a **readout** of this for future reference.*

when the modifier contains a number

*This relates to a standard **three-pin** plug.*

> When the number is a designation (where the second word forms a numbering system) a hyphen is not required.
>
> *We referred to this as a **phase 3 problem**.*

–en dash

The en dash is the length of a standard 'n' and is used mainly for notation to represent ranges and to split up names and opposites.

2001–2007   analog–digital   converter   input–output   pp.   123–134 phospho–L–serine

—em dash

The em dash is slightly longer than the hyphen and the en dash and has two functions. It is used at the end of a sentence for a final thought or a restatement of a previous thought. It is especially useful in long sentences when a set of commas has already been used. Only one dash is used for this task.

The algorithm behaves perfectly, especially for reconstructing the input signal, but the pitch shift is incapable of appearing in time-stretched signals—as observed in our design.

It can also be used in pairs instead of commas or parentheses to add emphasis and to clarify, or if the phrase interrupts the previous one. Parentheses should be used for interruption if the phrase carries little importance.

However, the low efficiency due to the wide band gaps in organic materials – usually around 2eV – is the main restriction to...

Do not use two en dashes or hyphens side by side for an em dash.   -- ✗

Parentheses

Parentheses are used in the main text to enclose a non-restrictive clause that adds information but has no real bearing on, or importance to, the sentence as a whole.

This cube is divided into many sub cubes (also known as voxels) that form...

The enclosed phrase does not start with a capital letter and any punctuation marks relating to the sentence as a whole should be outside the parentheses

…………(………………) . …………   ……….(…………...);………….

unless the phrase is a complete sentence standing on its own.

(…………………………………… .)

Punctuation marks for elements within the sentence remain within the parentheses.

……………(……,…… "……."………?)……………..

Apostrophes

The apostrophe is used to indicate both possession and contraction.

It is used to show that something belongs to a person or a group of people. Note that the apostrophe comes before the 's' for these singular nouns and after the 's' for the plural nouns.

company	*company's*	*companies'*
country	*country's*	*countries'*
user	*user's*	*users'*

its – The most important feature is it's (its) flexibility.   it's = (it is)

In this scenario the **users'** *devices will not be able to detect the threat level. (plural: more than one user)*

This could affect the **user's** *ability to detect a threat.* (singular: one user)

Some writers use the possessive apostrophe for inanimate objects, while others argue the of-phrase is the only acceptable form. A compound form with a generic meaning and without an apostrophe or an of-phrase is also starting to gain favor.

We will also look at the device's configuration.

We will also look at the configuration of the device.

We will also look at device configuration.

Regardless of preference, the first option should not be completely avoided – especially when the sentence is not generic and a specific entity is being referred to.

This system's performance *has been affected...[45] relates this to a cluttered* **system registry**.

Here a specific system is being referred to so the apostrophe form is appropriate.

Now we have the more general reference to 'system registry' and the generic third form is appropriate.

If there is no possession there should be no possessive apostrophe.

It will help designer's reuse many of their older versions.   ✗

The total number of subscriber's is 150.   ✗

Contractions are discouraged in academic writing.

We don't (do not) test for this until stage five is complete.

And apostrophes are not required for abbreviations or for dates.

SLA's   **SLAs**   1990's   **1990s**

SECTION III

Data and referencing

CHAPTER 21

Figures and tables

Introduction

Illustrations found in engineering papers can be categorised as either tables or figures. Photos, charts, models and graphs are all referred to as figures.

Graph 2. Frequency graph showing global and local resonances. ✗

Fig. 2. Frequency graph showing global and local resonances. ✓

Illustrations are not just useful for breaking up the text. A good illustration will provide the reader with the necessary information to assess the data collected or judge the argument that has gone before or that follows. Well-designed, informative and modern illustrations can also furnish a paper with a professional look and present the author as a capable and forward-thinking researcher.

Using figures

The first thing of note is that every reference to a figure in the text must be abbreviated to Fig. followed by the figure number.

Title: Fig. 2.2. Proposed switchboard modification

In text: *The proposed switchboard modification in Fig. 2.2. allows the...*

Fig. 2.2. shows the proposed switchboard modification that will...

The desirable size of a figure is less than one page. If the figure must overrun onto the next page then the full title should be written on both pages, i.e. repeated on the second page with any relevant information for that section.

first page   second page

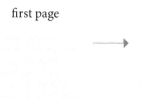

Fig. 1.6. (*Continued..*) Absorptance of organic solar cell in 3D simulation. (b) With grating depth of 150 nm.

Fig. 1.6. Absorptance of organic solar cell in 3D simulation. (a) With grating depth of 125 nm.

When mentioning two or more figures in the same sentence, name each of them separately to allow for cross-referencing:

see Fig. 5., Fig. 6., and Fig. 7.   ✓
see Figs. 5. through 7.   ✗

—axes

Axes on figures should be labeled using words and not symbols. Unit symbols can be placed in parentheses after the description.

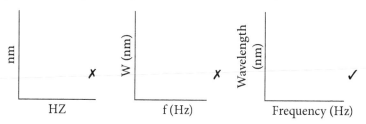

Using tables

Tables are numbered in roman numerals and the term 'TABLE' is written in capital letters. This appears above the actual table with the title one line below. The title should be in capital letters, centred and with diminishing line length.

TABLE IV

VARIABLES FOR LCL VSC CONVERTER LOSS CALCULATION.
(A) PEAK CURRENT. (B) PWM MODULATION INDEX.
(C) COSINE PHASE ANGLE

inverted pyramid
line length

No period at the
end

Permissions

If a figure is being used from another source then this should be referenced using a citation number and then included in the reference list at the end. Usually the copyright holder will provide the necessary text; if not use the following template:

<material> reprinted from <owner of copyright, title of publication, year of publication.>

When modifying a diagram from another source, even if significant changes have been made, it is always appropriate to acknowledge the work it is based on. Use 'adapted from' not 'adopted from' for this.

Fig. 12. The schematic of the patch clamping method (adapted from [23])

The grammar of figures and tables

—titles

Articles are normally excluded from the titles of figures and tables:

Fig. 4.5. A flow chart showing the cable selection process (second stage) ✗

Fig. 4.5. Flow chart showing cable selection process (second stage) ✓

The title should provide sufficient information for the reader to be able to interpret and understand the figure, including what, where and when, in the fewest words possible.

Fig. 1.8. Threshold values derived from transmission data ✗

Fig. 1.8. Example sets of uncompressed and compressed transmission data to determine threshold value ✓

—common errors

Note the article, tense, prepositional and adverbial errors that are made when referring to tables and figures.

> Articles are not used when the figure or table is numbered.

Some of these features are illustrated in ~~the~~ Fig. 3.

*The RAM is a 6 by 5 matrix as ~~below shown~~ (**shown below**).*

> 'below' should not be used as an adjective. Note that above can be used as an adjective: 'above table' / 'table above'

*The ~~below table~~ (**table below**) demonstrates the relationship between propagation time and receiver power.*

*Table. 2.2 ~~compared~~ (**compares**) the dynamic power of the SRAM bitcells. The power is clearly…*

> The simple past should not be used for introducing tables and figures.

*The converter topology is ~~showed~~ (**shown**) in Fig. 7.*

> The past participle (shown) not the simple past form is required after the verb 'to be' here.

Figure 6.8 shows ~~that~~ the capacitor voltage of the LCL circuit.

*These operations are shown ~~at~~ (**in**) Fig. 15. and Fig. 16.*

Remember to include the verb:

> Use 'in' for tables/figures not 'on' or 'at'.

*…as (**shown in**) Fig. 2.1.*

CHAPTER 22

Numbers and units

Introduction

Numbers and units appear throughout an engineering paper and some conventions are obvious, others less so. Reviewing the following guidelines and analyzing the example errors should lead to proficiency in this often overlooked area.

SI units

SI units represent the international standard and should be used in all research. If other units are desired then they should be placed in brackets.

*...measuring **8.9 cm** (3.5 in) when the device is fitted.*

The following is a list of commonly used and misused SI units with their abbreviations:

alternating current	**ac**	~~AC~~
amplitude modulation	**AM**	~~am~~
decibel	**dB**	~~db~~
degree Celsius	**°C**	~~°c~~
direct current	**dc**	~~DC~~
electronvolt	**eV**	~~ev~~

frequency modulation	**FM**	fm
field-effect transistor	**FET**	Fet
gigahertz	**GHz**	gHz
inductance–capacitance	*LC*	LC
kilohertz	**kHz**	KHz
kilojoule	**kJ**	kj
megabit per second	**Mbps**	mb/ps
microgram	**μg**	μG
microhenry	**μH**	μh
nanowatt	**nW**	Nw
phase modulation	**PM**	pm
resistance capacitance	*RC*	RC

Prefixes

yotta	10^{24}	
tera	10^{12}	
giga	10^{9}	giga- giant
mega	10^{6}	mega- great
kilo	10^{3}	
hecto	10^{2}	
deka	10	
deci	10^{-2}	
centi	10^{-2}	
milli	10^{-3}	
micro	10^{-6}	micro- small
nano	10^{-9}	multi- many
pico	10^{-12}	tele- far
yocto	10^{-24}	trans- across

Punctuation and spacing

For numbers higher than a thousand, the units should be separated by a thin space rather than a comma – as a comma in some languages indicates a decimal point.

The variable range of deflection is from 15,000 to 35,000.   ✗

The variable range of deflection is from 15 000 to 35 000.   ✓

A thin space should also separate the number from the SI unit; compound units should be separated by a centre dot.

Signals from those sensors were acquired at a sampling frequency of **50 kHz.**

This has a density range from 1.5 to 1.9 **g·cm.**

Numbers and units do not need to be hyphenated unless they are being used as adjectives and the hyphen improves the clarity of the phrase.

The signal increases to 60-mV in 0.3-ms.   ✗

The signal increases to 60 mV in 0.3 ms.   ✓

The electrical field intensity had a **10-kV voltage** *at this height.*

Style

- A zero should always be placed before the decimal point but is not required after.

  ~~.41~~   **0.41**   ~~1.10~~   **1.1**

- Using a number and a unit to modify a scale should be avoided where possible and rewritten with an 'of' phrase instead.

  This resulted in a 72.5 mm height.   ✗

  This resulted in a height of 72.5 mm.   ✓

- Numbers (not words) should always be used with units.

  *The values lag behind those of the SOLA approach by approximately ~~twelve~~ (**12**) dB.*

- If the number is not linked to a unit then spell the word out for numbers below 11, unless they are included in a range.

  *We developed **seven** different structures by optimization.*

  *An ideal range would be from ~~eight to sixteen~~ (**8 to 16**).*

- And use numbers for 11 and above unless they begin a sentence.

  *In [32], the team discovered ~~three hundred and seventy nine~~ (**379**).*

  *6 (**Six**) should be enough to cover all the areas in our testing region.*

- Units that are not linked to numbers should be spelled out.

  *Naturally this would also be measured in ~~Hz~~ (**hertz**).*

- Units of measure should be pluralised unless the quantity equals one or is part of a compound term:

  *[22] suggest changing to **32 bits** to increase the support.*

  *We would recommend **16 bit AS numbers** for this part.*

Position of the number

Numbers go after the sequence word (first, next, last) and before the adjective when modifying a noun.

*The **next three lines** display the peers for each router.*

*These are the **two main contributions** of this study.*

*We illustrate here the **twenty successive subcarriers** that…*

We add additional three links to this ontology.   ✗

We add three additional links to this ontology.   ✓

Percentages

Use 'percent' or % but not 'percentage/percentages' with numbers.

The variation is slightly higher than 80 ~~percentage~~ *(%/percent).*

You can write 'percentages' when it is not connected to an amount.

The **percentages** *for desired hits and received hits are 77 and 56 respectively.*

Remember the 'of' phrase in this construction:

28% **(of the)** *files were images.*

And do not use a comma as a decimal point.

~~98,9%~~   **(98.9%)**

Errors

The ratio is between ~~0 to 1~~ **(0 and 1).**   *The levels will vary from* ~~0 and 5~~ **(0 to 5).**

This is likely to be ~~on~~ **(in)** *the range of 70–150.*

As a result, there are 100 ~~number of~~ *points in the wavelength component.*

All four ~~numbers of~~ *control valves are responsible for...*

We have found ~~numbers of~~ **(many)** *attributes including better gate controllability...*

It is then increased from 3.7 kbps ~~and~~ **(to)** *13.1 kbps.*

This will not be measured if it is below ~~than~~ *2.*

The user must select a number over ~~than~~ *5.*

~~A further number of~~ **(Additional)** *unipolar devices will be required for...*

This amounted to three ~~hundreds~~ **(hundred)** *components.*

We estimated the cycles to be around eight ~~thousands~~ **(thousand).**

> Hundred and thousand should be in singular form if attached to a number.

A case was made for two distributed antenna ~~system~~ **(systems).**

Cardinal, ordinal and fractions

~~thirtey~~ **thirty (30)**	~~fourty~~ **forty**	~~fifthty~~ **fifty (50)**
~~eightey~~ **eighty (80)**	~~eight one~~ **eighty one (81)**	~~two hundreds~~ **two hundred (200)**
~~forth~~ **fourth (4th)**	~~twelth twelf~~ **twelfth (12th)**	~~fourtieth fortyth~~ **fortieth (40th)**
~~two third~~ **two thirds**	~~one quarters~~ **one quarter**	~~third quarters~~ **three quarters**

CHAPTER 23

Equations

Introduction

Mathematics is an essential part of engineering. Translating problems into mathematical expressions and then into understandable English requires patience, and an awareness of the simplest way to project your methods and processes so the reader can best understand them.

Use simple terms ✓
Define everything ✓
Check for sign errors ✓
Check the order of your operations ✓

This section provides the correct notation and English expressions alongside the key errors made by engineering writers when adopting this specialized form of English.

Formatting

—numbering

Equations need to be numbered uniquely and consecutively in the text.

Different versions are accepted as long as consistency is maintained:

(1a) ✓   (1.1) ✓   (1–1) ✓   [1.1] ✗   (1,1) ✗   (1~1) ✗   (1:1) ✗

The reference number should be right aligned and in round brackets:

$$a + b = c \qquad\qquad (1)$$

—punctuation

When equations are embedded into the text try to keep them on one line, using brackets and a forward slash for division:

$$v = \frac{c}{n} \text{ becomes } v = c/n \qquad \frac{c}{b-n} \text{ becomes } c/(b-n)$$

Colons are not used to introduce an equation unless the text is a complete sentence.

We can see that

$$y(t) = \frac{1}{N} \sum_{i=0}^{N-1} x(t-i)$$

The moving average filter is as follows: ⟵ 'following' and 'as follows' always take a colon.

$$y(t) = \frac{1}{N} \sum_{i=0}^{N-1} x(t-i)$$

Do not use a comma after forms of the verb 'to be'.

this is   the result being

But use a comma after

thus, hence, i.e., e.g.,

Ellipses always contain three dots and should be surrounded by commas.

$E = 1, 2, 3,..., n$

Equations should be referenced by their number only and in round brackets, unless they are sentence starters:

Observe that in equation (2–32) the control signal is... ✗

Observe that in eq. (2–32) the control signal is... ✗

Observe that in (2–32) the control signal is... ✓

Equation (2–32) includes the conduction and switching losses... ✓

Check your use of the following pairs and groups:

Bracket order: $\{[()]\}$

round brackets ( )	square brackets [ ]	braces { }
angle brackets ⟨⟩	greater than/less than > <	
superscript ^{2}	subscript $_2$	
zero 0	letter o	

—breaks

Equations are normally broken at the operator and set right aligned.

$$x = (5\alpha + x)$$
$$- (5y - \alpha + z)^2$$

Unless the verb occurs on the right hand side of the equation, then

$$5\alpha + x + 5y$$
$$+ \alpha^2 + z = x$$

Terminology

The language of mathematics is specialized but most writers adapt to the style well. The most important aspects are explaining each step to the reader using relevant and accurate terms, and defining each element beforehand or immediately afterwards.

A few combinations	A few prepositional terms	
Let...be	applied to	denoted by
Applying...gives	derived from	equivalent to
Interchanging...yields	expressed by	formed by
Reorganizing...results in	holding for	indicated by
Suppose...then	represented by	specified by

As the boxes illustrate, gerunds are often employed to begin sentences and terms are frequently partnered with the preposition 'by'.

—the equals sign

In mathematics, the equals sign (=) is considered a verb and is the fundamental part of an equation. Without an equals sign the information cannot be referred to as an equation. It is an expression instead.

*The following ~~equation~~ (**expression**) is our initial building block:*
(x + 2)(y − 5)

Avoid using the equals sign in the main text. It should only be used in an equation.

The total inputs are = to the sum of the individual inputs.   ✗

And Pp = the eventual product when...   ✗

This is also true of the greater than/less than signs if not part of an expression.

The phase difference should be > the minimum phase difference...   ✗

Further suggestions

- Use verbs in an active voice and avoid nouns and awkward 'of' phrases:

  It can be observed that...   ✗

  We observe that...   ✓

  The substitution of (3.8) will lead to the expression of...   ✗

  Substituting (3.8), we can express   ✓

- Make sure that the language is parallel.

  ignoring....and assuming...

  applying....and using...we can rewrite..

  *Finally, we divide...and ~~calculating~~ (**calculate**)...*

- Make sure that all symbols are defined, either before or immediately after equations.

$$\overline{V_{1ac}} = V_{1acm} \angle \alpha_1 = V_{1acd} + jV_{1acq} \tag{3-5}$$

$$\overline{V_{2ac}} = V_{2acm} \angle \alpha_2 = V_{2acd} + jV_{2acq} \tag{3-6}$$

where, $\overline{V_{1ac}}$, $\overline{V_{2ac}}$ are the voltage phasors, V_{1acm}, V_{2acm} are the magnitudes, α_1, α_2 are phase angles...

- Use ***at*** *time t* not ~~*in*~~ *time t.*
- Use 'gives', 'produces' or 'yields', not 'gets'.

  Following the previous algorithm, the choice of (j-2) ~~*gets*~~ ***(yields)****:*
- Check whether you require 'an' or 'any'

  Let x be an element of x...   Let x be any element of x...
- Objects or expressions should never be referred to as 'it' or 'this' or 'they'. Always name the object or expression for clarity and accuracy. Do not worry about repetition—just make sure the reader can follow the method.

  Based on (6b), it is calculated as: ✗   *Based on (6b), the angle* β_2 *is calculated as:* ✓
- Give as much information as possible to the reader.

  From (3) it follows that   *From (3) it follows that* ***if***
                                       $V(sk) > 0$ ***then***

                   ✗   ✓

  $V(k) = V(sk)V(B)$ *(4)* $V(k) = V(sk)V(B)$ *(4)*

Take care

Think about the connecting terms you are using and whether they are actually suitable for what follows.

When using thus/hence/so...does it follow or does it actually contradict or require further explanation?

When using otherwise/else...is it actually a contradictory or alternative outcome?

When using similarly/correspondingly...is it actually similar and does it correspond?

When using reconstructing/reapplying...are you in fact just constructing and applying?

Example errors

*Let μ and G(s) ~~is~~ (**be**) the mean and the…*

*Where xa(μ) and ya(μ) ~~be~~ (**are**) the coordinates of the transformed shape.*

> Note the different forms of the verb 'to be' with let, suppose and where.

*Suppose the 3D markers ~~be~~ (**are**) feature vectors…*

*We can express X ~~to be~~ (**as**) the union of two…*

*The length ~~equals to~~ (**equals**) the number of nodes of the sub-network.*

> Do not add 'to' if you are using 'equals'. You can write 'is equal to'.

*~~Rearrangeing~~ (**Rearranging**) (3.2) gives…*

*Our model calculates the correction factor that is ~~summed to~~ (**added to**) the ANN output.*

*The dielectric constant is ~~given with~~ (**given by**) $\varepsilon = \varepsilon_0(1 + \chi)$.*

*Where P(C_1) and P(C_2) ~~is~~ (**are**) the probability…*

*Equation (2.9) displays the loss probability as a function ~~for~~ (**of**) the effective arrival rate ($\lambda_1 + \lambda_2$).*

> Make sure the verb agrees with the number of elements.
>
> *Where [0, 1, 2, 3….K] are…* ✗
>
> *…(1.2) and (1.3) is…* ✗
>
> *This set of variables are…* ✗
>
> *If xA are transmitted…* ✗

*We can then ~~substract~~ (**subtract**) the newly inserted coefficients.*

CHAPTER 24

Referencing

Introduction

This section deals with referencing work in the main text and also in the reference list (bibliography). Correct referencing requires patience and precision and crucially a set of samples against which you can check the accuracy of your work. These templates are provided here.

The number system

Engineering papers use the number system to cite work in the main text.

This has also been applied to information systems [5].

Grammatically, this numbering system can be thought of as a direct replacement for the author's name.

In fact, [5] analysed the drawbacks of this tool as well.

The numbering can also be used in an indirect manner, in the style of footnotes.

The key features of this tool have been analysed [5].

The names of authors can appear in the text if the nature of the sentence requires them to be stated.

Morgan met with the chief engineer to query this [5].

If you wish to guide the reader to specific chapters, figures or equations in a work then use the following system:

…which details the design principles of this convertor [5, Ch. 3].

The switching currents have also been modeled [5, Fig. 6].

These three variables can also be switched [5, eq. 9].

—a few example errors...

Enhancing the security of this grid infrastructure is therefore considered vital [see 5].

> Text should not be placed within the brackets

We also find this system being used by Morgan and Jones [5].

> Use square brackets not round ones.

> Names are not required in the text.

Taking a different view, (5) [5] stresses that invalid switching is a common cause.

> ibid should not be used. Just use one number for each paper and reference it.

The conduction voltage for this has already been estimated [ibid.].

Reference list (bibliography)

Before looking at each type of work in detail, here are a few conventions to follow in the reference list.

Regardless of the type of work, the initials of the author's given names are followed by the family name.   S.N. Morgan

If there are six or more authors then *et al.* can be used.   S.N. Morgan *et al.*

—abbreviations

General:

p.	for one page p.5		ed.	edition
pp.	for more than one page pp.3–9		Ed.	editor
vol.	volume		ch.	chapter
no.	number		sec.	section

Titles:

Inst.	institute	Sci	science
Int.	international	Technol.	technology
Org.	organization	Ann.	annals
Soc.	society	Lett.	letters
Nat.	national	Mag.	magazine
Univ.	university	Netw.	networks/networking
Proc.	proceedings	Rec.	record
Trans.	transactions	Sel.	selected
Syst.	system	Commun.	communications
Autom.	automatic/ automation	Telecommun.	telecommunications
Bus.	business	Intell.	intelligent
Comput.	computer/ computational	Spectr.	spectrum
Eng.	engineering	Amer.	America
Ind.	industry/ industrial	Chin.	Chinese
Manuf.	manufacturing	Jpn.	Japan

—books

S. Morgan, *Electrical Engineering*. New York, NY, USA: Wiley, 2009.

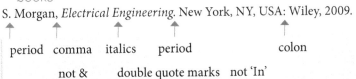

period	comma	italics	period	colon
	not &	double quote marks	not 'In'	

S. Morgan and R. Smith, "Power networks," in *Electrical Engineering*, P .L. Jones, Ed. Washington, DC, USA: NBS, 2009, pp. 123–145.

—periodicals

S. Morgan, "Surface antennas," *IEEE Trans. Wireless Commun.*, vol. 61, no.9, pp. 230–240, Feb. 1999.

| date | Initial capital on first word only | italics | abbreviated forms |

journal acronym (pre-1988)   double quote marks

S. Morgan, R. Smith, and P. L. Jones, "Optimal design," *IEEE Trans. Aerosp. Electron. Syst.*, vol. AES-4, no. 5, pp. 129–135, Apr. 1967.

—conference proceedings

S. Morgan, "Transient signal propagation," in *Conf. Rec. 1999 IEEE Int. Conf. Commun.*, pp. 22–34.

no need for date at the end if already in the reference   use abbreviations and omit articles and prepositions from the conference title

—dissertations and theses

S. Morgan, "Surface-wave radiation," Ph.D. dissertation, Dept. Elect. Eng., Harvard Univ., Cambridge, MA, 2010.

—online resources

S. Morgan, "Nano-structured surfaces", Chicago, IL, USA: Univ. Chicago Press, 2013. [Online]. Available: http://www.elecweb.com.

square brackets   not 'available at'

break a URL after a slash and before a hyphen   not 'assessed'

The Antenna eBook. Rengall Corp., 2011. [Online]. Available: http://rengal.com/eBook/antenna_ebook_pdf_1945_sr.pdf. Accessed on: May 5, 2012.

A–Z list of errors

A–Z list of errors

absorption/adsorption

Take care with these three similar looking nouns:

absorption – conversion or interception of energy or waves
absorptance – the effectiveness of absorbing radiant energy
adsorption – the binding of molecules to a surface

adsorption occurs on the surface; absorption occurs throughout.

*High rate **absorption** will occur in the active layer to enhance the **absorptance** of the organic solar cell.*

*This relates to the **adsorption** of ions in microporous materials.*

Verb forms: absorb/adsorb

access

Remember to include both double letters (cc, ss)

This spelling mistake also occurs with 'assess' and 'possess', but note the spelling of 'recesses'.

*Then each user ~~acceses~~ (**accesses**) the subcarriers.*

*Static random ~~acess~~ (**access**) memory (SRAM) is used as a memory cache.*

accumulator

This is usually a spelling error but it is prolific.

*It then stores this in the ~~acculminator~~ (**accumulator**) until the desired time.*

The error may be explained by the verb form 'accumulate' being confused with the similar sounding verb 'culminate' (to end in).

*Their efforts ~~acculminated~~ (**culminated**) in a much more efficient system.*

Also, 'accumulator' will require an article if it is being used in the singular form.

*We will load the data from (**the**) accumulator to the other register.*

adjust/adjustment

adjust – verb

adjustment – noun

*They made the necessary ~~adjust~~ (**adjustment**) in [12].*

*After several ~~adjusts~~ (**adjustments**), the frequency weighting functions were determined.*

advantage/benefit

Although both have noun and verb forms, the verb form of advantage is rare and benefit is the likely option.

*They will ~~advantage~~ (**benefit**) from this procedure in the long run.*

'advantage' is used in the phrase 'take advantage of', meaning to make use of certain conditions for your own gain.

*These members are likely to **take advantage of** this situation and…*

adverse

'adversary' is a noun meaning an opponent.

The adjective form 'adverse' meaning unfavorable or opposite is normally required.

*We explored this ~~adversary~~ (**adverse**) impact in the previous chapter.*

The adverb form is 'adversely'.

*If we stop the threat now then the user will not be **adversely** affected.*

algorithm

Simply a spelling error, but keep an eye on the awkward ending.

This is also possible with a novel control ~~algorithum~~ (**algorithm**) *[22].*

The grid protection ~~algorithmn~~ (**algorithm**) *was applied to the system.*

also/as well

'as well' normally goes at the end of a sentence.

This component could play an important part **as well.**

It can only go at the beginning in the form 'As well as...'

As well *as reducing stability, it* **also** *affected the system performance.*

'as well' is always two words.   ~~aswell~~

'Also' can appear at the beginning of a clause as a conjunctive adverb.

*...~~as well~~, (**also**,) there was very little funding available at the time.*

Or be used as an adverb modifying verbs and adjectives.

*This could ~~as well~~ (**also**) affect the original parameters.*

ambient

ambient – adjective

ambience – noun (the atmosphere of a place)

'ambient' normally describes temperature, light, air or noise.

ΔT *is the* ~~ambience~~ (**ambient**) *temperature rise and K is the constant.*

amplify

Note the endings of the verb forms and also the noun.

amplify – verb

This will amplify the low level signal.

It has amplified the internal noise...

*Next we will look at the amplif**ying** matrices.*

amplification – noun

*This is the standard amplif**ication** method.*

analyse/analysis

analyse – verb
analysis – noun   (plural: analyses)

*We will ~~analysis~~ (**analyse**) the consequences of this low voltage.*

*Finally, there is an ~~analyse~~ (**analysis**) of the incremental cost.*

Remember to use the plural form of the noun when required.

*Many ~~analysis~~ (**analyses**) have failed to consider the supporting circuits.*

The preposition 'of' is used after the noun.

*There will also be an ~~analysis about~~ (**analysis of**) evolving networks.*

arbitrary

arbitrary   adjective – not having complete relevance, only relative.

The associated adverb is 'arbitrarily'.

In mathematics, something that is arbitrary has no specific value.

*We can get the electrical field intensity value at any **arbitrary** point and the electrical field distribution in any **arbitrary** area by FEM calculation.*

*If we use this sensor it can be mounted **arbitrarily**…*

as is the case

This is a fixed phrase. Do not add any extra words.

This can be carried out immediately, as ~~it~~ is the case with other routing updates…

authentic

A prolific word in engineering; here are the main forms and examples of use:

authentic – adjective
authenticate – verb

*This will reveal whether it is **authentic** or malicious.*

*We need to **authenticate** this as soon as possible.*

A distinction can be made between the two associated nouns.

'authenticity' relates to the quality of being genuine or authentic; 'authentication' is the process of identifying whether something is authentic or genuine.

*They also looked at the **authenticity** of the information exchanged.*

*This provides **authentication** of many of the nodes.*

automaton

automaton – a self-operating machine or device, traditionally a mechanical copy of reality. ~~automatum~~

The plural is 'automata'. *This is an example of finite state **automata** (FSA).*

'automation' can be used for an electronic device but is more commonly employed for the automatic control of a system e.g. by IEDs or data from remote sources.

*We saw in [9] how they simulated the dynamics using a two-dimensional cellular **automaton**…*

*The standards relate to the **automation** of data acquisition and protection.*

auxiliary

A horrible one to spell: 'auxiliary' has no double letters, is an adjective and describes something in an assisting or supporting role. It frequently describes transmitters and transformers.

*The station provides an ~~auxillary~~ (**auxiliary**) transformer for the LV network.*

-band

Terms that contain the suffix –band are normally written as one word:

baseband   broadband   narrowband   passband   wideband

*The next section will present the **narrowband** fading analysis.*

But 'dual band' is written as two words.

*In [18] a **dual band** compressor was employed for this task.*

Be careful with the term 'wide band gap' as the band part is linked to 'gap' not 'wide'.

*These semiconductors are ~~wideband gap~~ (**wide band gap**) and allow the devices to operate at higher temperatures.*

bandwidth

The noun 'bandwidth' is one word.

*This shift is proportional to the system ~~band width~~ (**bandwidth**).*

Also check the spelling:

~~bandwith~~

basic/basis

basic adjective – essential; underlying
basis noun – the main principle, the foundation

*The models may be used on an application-specific ~~basic~~ (**basis**).*

*This is a ~~basis~~ (**basic**) principle of this system.*

The plural of 'basis' is 'bases'.

*There are actually several **bases** on which to develop these security measures.*

beside/besides

beside preposition – near; by the side of

*The equipment to be used was placed **beside** each computer.*

besides adverb – furthermore, also

Besides, the process is already moving towards nanometer technology.

preposition – other than; in addition to

~~Beside~~ (**Besides**) improving the SRAM at circuit-level…

bottleneck

Always one word:

Although the traffic had eased at this point, there were still considerable ~~bottle necks~~ (**bottlenecks**) *in other areas.*

brake/break

Check your use of these two homophones:

break – an interruption or pause.
brake – a device for slowing or stopping a mechanism.

This is known as a regenerative ~~breaking~~ (**braking**) *system.*

Obviously demand will fall during these ~~brakes~~ (**breaks**).

The circuit ~~braker~~ (**breaker**) *is tripped before any…*

bypass

Always one word, as is the case with 'bystander'.

…where [11] and [12] have evaluated a number of **bypass** *capacitors.*

byte

bit – the smallest unit of storage
byte – a grouping of eight bits

This is also measured in ~~kilobites~~ (**kilobytes**).

The resolution of the sensor can be configured to 10, 11, or 12 ~~bytes~~ (**bits**).

cache

Check the spelling and do not capitalize.

*We have to minimize the ~~cashe~~ (**cache**) vulnerability here.*

*The microprocessor will then write to the ~~Cache~~ (**cache**).*

called

Do not add 'as' when you are stating what something is called.

This was later called ~~as~~ the optimal input problem.

The same applies to 'named' and 'termed'.

They looked at a connection named ~~as~~ Mission Farm PV generation (MFPV).

Perhaps the confusion stems from the term '*known as*'.

capacitor/capacitance

capacitor – A two-terminal electrical component

capacitance – A measure of the capacity of storing electric charge

*The energy is then stored in this ~~capacitance~~ (**capacitor**).*

*These represent resistance, inductance and ~~capacitor~~ (**capacitance**) respectively.*

carry out

'Carried out' is preferable to 'done'.

*Further work will be ~~done~~ (**carried out**) to automate the network elements…*

Remember to include the second part of the phrase.

*More investigation will be carried (**out**) to evaluate the performance of the algorithms.*

causal/casual

'casual', which means relaxed or unconcerned, is often confused with 'causal', which refers to a cause.

*We will aim to establish a ~~casual~~ (**causal**) relationship between the variables.*

The noun 'causality' is the relationship between cause and effect.

choice/choose

choose verb – to select or decide (past tense: chose

past participle: chosen)

choice noun – the act of choosing; a selection or alternative   ~~choise~~

*The engineer must then ~~chose~~ (**choose**) a suitable method.*

*They were asked to make a ~~choose~~ (**choice**) between addressing schemes.*

collude/collusion

collusion   noun – malicious coordinated behavior of a group of users

The verb is 'to collude'.

*This scheme seeks to prevent ~~colluding~~ (**collusion**) attacks in sensor networks.*

*These users will look to ~~collusion~~ (**collude**) at this stage of the process.*

combining

Often just a lack of concentration but nevertheless a common typo.

*~~Combing~~ (**Combining**) both variables with historic load data…*

commutate

commutate   verb – to reverse every other half cycle so as to form a unidirectional current

The noun form is 'commutation'.

*First, we introduce the integrated gate ~~communtated~~ (**commutated**) thyristor (IGCT).*

*Whereas, [11] provided **commutation** analysis of single phase motors.*

compete/complete

This is another instance where the similarity in spelling is the likely cause:

*Another possible way of achieving a ~~compete~~ (**complete**) specification is…*

*In this case they would be ~~completing~~ (**competing**) for the same power supply.*

complement/compliment

complement   noun – something that adds to or completes something else

compliment   noun – an expression of praise

Both have verb forms identical to the noun: 'to complement' and 'to compliment'.

*The next section discusses how these two processes ~~compliment~~ (**complement**) each other.*

*This is the greatest ~~complement~~ (**compliment**) we can pay to this system.*

complied/compiled

comply – to be in accordance with

compile – to put together or gather

The error normally occurs with the past tense of these verbs.

*Their chips ~~compiled~~ (**complied**) with the IEEE 802.15.4 physical layer standard.*

*The binaries are then ~~complied~~ (**compiled**) for each hardware platform.*

The adjective form of comply is 'compliant'.

*This is **compliant** with the IEEE 802.15.4–2006 standard.*

compute/computation

compute – verb

computation – noun

A multilevel network should reduce the ~~compute~~ (**computation**) *time substantially.*

The next step is to **compute** *the six features.*

The associated adjective for compute is 'computable'.

The associated adjective for computation is 'computational'.

concern/consider

'concern' is one of the most misused verbs in English.

verb – to relate to

This section **concerns** *the problems with the network charge.*

verb – to trouble or worry

They were especially **concerned** *about the load densities.*

The following situations require 'consider' not 'concern':

Morgan and colleagues ~~concern~~ (**consider**) *such a relationship in [11] and [12].*

Both technical and economic aspects were ~~concerned~~ (**considered**) *when analyzing this.*

confidence

This entry relates principally to the mathematical phrases 'confidence value' and 'confidence interval' where the adjective form is inserted by mistake.

The obtained associations and the ~~confident~~ (**confidence**) *values are detailed next.*

configured

The opposite of configured is generally considered to be 'unconfigured'.

'misconfigured' is widely used jargon for not configured, while 'disconfigured' is used for a change to the appearance of something.

The o-state field can be configured or **unconfigured.**

constrain/constraint

constrain   verb – to repress or confine

constraint   noun – limitation or restriction

Note that the plural noun is 'constraints' not 'constrains'.

*It is important to recognize any design constrains (**constraints**).*

Using the noun instead of the verb is common.

*These barriers will constraint (**constrain**) the transconductance.*

consume/consumption

consume – verb

consumption – noun

*It is unsuitable for most sensor architectures due to high energy consuming (**consumption**).*

*This will require too much memory space and will **consume** a lot of resources when processing the signal.*

contiguous

A frequent misspelling:

contigous   contiguos   cotiguous

*It is possible to use a **contiguous** network block here.*

Networks can also be described as discontiguous and noncontiguous. The two are similar in meaning, but one distinction is that discontiguous implies separation whereas noncontiguous indicates absence.

continual/continuous

There is a slight difference between these two adjectives.

continual – starting and stopping on an interval basis

continuous – non-stop or never-ending; without interruption

*The **continual** problems were very frustrating for the service users.*

*This drives the DUT with a **continuous** wave signal.*

Sometimes 'continues' is written by mistake.

contrary

~~In the contrary~~	Use **On the contrary**
~~In the contrast~~	Use **In contrast**
~~In contrary to~~	Use **Contrary to**

On the contrary, the first study did not even consider the oversampling factor.

In contrast, the LPC2478 is able to meet these requirements.

Contrary to the findings in [32] we found no instances at all.

There is little general difference between the phrases, but 'in contrast' is normally used as a mere comparison whereas the other two ('on the contrary'/'contrary to') are used to clearly state the opposite and show disagreement.

'On the contrary' can only be used as a response to something just mentioned.

~~On the contrary~~ to the findings in [3]...

*The findings are not disappointing as [11] and [14] claim; **on the contrary**, they provide a number of interesting...*

coordinate

Can be a noun, adjective or verb; a hyphen is not required but spelling is an issue.

~~cordinate~~   ~~coordinant~~   ~~coordernate~~

*They looked at general **coordinates** and b-field gauge transformations.* (noun)

*We need to **coordinate** it so all users are on the network at the same time.* (verb)

crosstalk

'crosstalk' is an uncountable noun and is always one word. Do not abbreviate it or use a plural form.

~~cross talk~~   ~~xtalk~~   ~~crosstalks~~

*These signals and wavelength channels influence each other and create significant ~~cross talks~~ (**crosstalk**).*

cycle/cyclical

There are two adjective forms of the noun 'cycle' with no discernible difference in meaning.

*This led to a **cyclic** dependency between the users.*

*…[12] noted some **cyclical** stressing of the solder joints.*

The adverb form 'cyclically' is awkward to say and often overlooked in writing. It is required in the following instance instead of the adjective:

*We use it to ~~cyclical~~ (**cyclically**) display the process parameters in the first line.*

Cyclic is always used for the term 'cyclic redundancy'. ~~cyclical redundancy~~

daemon

Check the spelling and do not capitalize this term meaning a dormant program lying in wait.

~~Daemon~~   ~~demon~~

*As with all ~~deamons~~ (**daemons**) this will continue to run in the background.*

deduce/deduct

deduce   verb – to reach a conclusion through reasoning; to infer
deduct   verb – to take away

*If we ~~deduce~~ (**deduct**) the corresponding row we are left with…*

*They ~~deducted~~ (**deduced**) that this was primarily being caused by…*

defective

The noun and adjective forms are linked.

defect   noun – a fault or imperfection

defective   adjective – faulty; imperfect

But the verb is unrelated and 'defected' is not used in engineering.

*The next step is to isolate the ~~defected~~ (**defective**) equipment.*

detect/detection

detect – verb

detection – noun

*We are interested in the ~~detect~~ (**detection**) and classification of faults.*

*This power is sufficient enough to **detect** these line strengths.*

The electrical component ends in –or.

*A millimeter wave ~~detecter~~ (**detector**) was then attached to the cavity.*

device/devise

device   noun – a machine or tool used for a specific task

devise   verb – to form a plan or invent

*Insulation overload could have affected the electrical ~~devise~~ (**device**).*

diameter

This spelling error appears from time to time.

*The ~~diemeter~~ (**diameter**) of the pipe should be no more than 250 mm.*

dielectric

Dielectric can be used as a noun or an adjective. Remember to retain the initial 'e' of electric.

*Whereas [11] and [12] concentrated on insulation layer thickness and ~~dilectric~~ (**dielectric**) properties…*

*The **dielectric** here is surrounded by a metal gate.*

disconnect

'Disconnect' can be used as a noun in electronics but this use has not been extended to 'connect'.

*The objective is to build it without having a **disconnect** between the modeling and the experiments.*

*A ~~connect~~ (**connection**) can be made through the interface.*

dissipation

The error here is straightforward verb/noun confusion.

dissipation – noun

dissipate – verb

*This represents less than 2% of the total static power ~~dissipate~~ (**dissipation**).*

*The power **dissipated** at the load would be much more than this.*

distortion

Be careful with the spelling.

*All of these can cause ~~distorsion~~ (**distortion**) to the EM wave.*

disturb/disturbance

disturb – verb

disturbance – noun

*The attacker will hope to **disturb** the session with this maneuver.*

*This measure will improve the distribution and reduce the ~~disturbs~~ (**disturbances**).*

dominant

dominant – adjective

dominance – noun

*These are the ~~dominance~~ (**dominant**) nodes and will affect system performance. This **dominance** can be reduced by...*

*...which is referring to the active or ~~dominance~~ (**dominant**) power supply.*

dynamic

Used as a noun and an adjective. The plural noun is 'dynamics'.

A distinction can be made with the related adjective 'dynamical':

dynamic – relating to change; nonstatic; requiring periodic attention.

dynamical – relating to dynamics in general.

'Dynamic' is the more common term for describing processes and methods. Check the relevant literature when using specific terms. A few are listed below:

dynamic braking   dynamical linear...finite-dimensional system
dynamic random access memory   nonlinear dynamical systems
dynamic scheduling   dynamic voltage scaling

eavesdrop

The term is always one word.

*It would be better to (eaves drop) **eavesdrop** on them as they have yet to register.*

effect/affect

affect   verb – to have an influence on
effect   verb – to accomplish

*This may effect (**affect**) the system in the long term.*

*This affected (**effected**) the change to wireless technology.*

More commonly, 'effect' is a noun meaning result or consequence.

*This had no affect (**effect**) on the results.*

efficient/efficiency

efficient – adjective
efficiency – noun

*Having these kinds of routes ensures the efficiency (**efficient**) use of bandwidth.*

In this case, high ~~efficient~~ (***efficiency***) *can be achieved by designing a scheme that can send* ***efficient*** *economic signals.*

electric/electrical

There is great similarity between the two adjectives but a distinction can be made:

electric – usually used for things that run on electricity

electrical – usually used for things relating to electricity in general including occupations and fields

The term 'electronic' often pertains to equipment with small electrical parts. Electronic entities use the properties of electrons for information whereas electric/electrical entities use electricity as energy for power.

Possibly because of the figurative use of electric the term 'electrical' is preferred in some instances. It is best to check individual terms.

electrolyte

'Electrolysis' is the process, 'electrolyte' is the substance or medium and both are nouns. 'Electrolytic' is the related adjective.

~~electrolite~~   ~~electrolises~~

There are three major components: the ~~electrolytic~~ (***electrolyte***), *cathode, and anode.*

emit/emission

emit   verb

emission   noun – something that is emitted; energy emitted from a source

emittance   noun – the energy radiated by the surface of a body

The radar signal should ~~emission~~ (***emit***) *in short electromagnetic pulses.*

This fuzzy logic system is used as a powerful tool to model the acoustic ~~emit~~ (***emission***) *signal.*

They considered different ways of reducing the infrared ***emittance***.

encrypt

Note the noun in the first example and the different forms of the verb 'to encrypt' with their associated errors:

*This is an algorithm for performing an ~~encrypt~~ (**encryption**).*
*This is the message in plain form before being ~~encripted~~ (**encrypted**).*
*They will be able (**to**) encrypt in this way.*
*The method is used for ~~encrypt~~ (**encrypting**) the data.*

exceed/excess

exceed   verb – to go beyond with regard to quantity, degree or rate.

excess   noun – an extreme degree; going beyond the limits.

*The problem is that this may ~~excess~~ (**exceed**) the limit.*

except for/apart from

'except for' excludes something or someone.

*They all have a complex refractive index **except for** the lossless materials.*

'apart from' can exclude OR include something or someone.

***Apart from** high core loss, the transformer would also be very heavy.* (include)

*The loopback networks have been installed **apart from** the former two.* (exclude)

The error is made when trying to use 'except for' to include something.

*~~Except for~~ (**Apart from**) grating period, the variables have yet to be assessed.*

Be careful not to confuse 'for' and 'from' in these terms.

~~except from~~............ ~~apart for~~

excite/exciton

excite   verb – to supply with electricity for producing electric activity

exciton   noun – a bound state of an electron and a hole

*In this cell photons can **excite** the electron in the valence…*

*It is a dynamo that is called a ~~self-exited~~ (**self-excited**) generator.*

*The main issue is the short ~~excite~~ (**exciton**) diffusion length.*

existing

'Existing' is an adjective with two associated issues. One is a typo or spelling error and the other is the unnecessary use of the adverb 'already'.

*From this, new nodes can be added and ~~exiting~~ (**existing**) links between nodes can be reallocated.*

*The next stage is to identify an ~~already~~ (**existing**) member of the network.*

expect/expectation

expect – verb

expectation – noun

*The absorptance curve is more or less similar to the ~~expect~~ (**expectation**).*

*As ~~expectation~~ (**expected**), the incurred overhead is slightly higher.*

extend/extent

'extend' is a verb meaning to stretch out or increase.

*We **extended** this to the boundary of the computation cell.*

'extent' is a noun meaning the degree to which something extends.

*The **extent** of this problem can be seen from the statistics below.*

The following error is commonly made:

*It is not really known to what ~~extend~~ (**extent**) the architecture is responsible.*

feedback

Feedback is evolving into a countable noun as well as an uncountable one. The noun is written as one word and the phrasal verb is two.

*We can then **feed** this information **back** into the system.*

*The carrier is temperature stabilized with a **feedback** loop.*

*This excludes all the natural interactions and **feedbacks.***

following/as follows

The different forms and the main errors are listed.

*...will be explained in the ~~follow~~ parts: **following***

*The advantages are described as ~~following~~: **follows***

*The ~~followings~~ are some suggestions: **following***

to follow – verb   to come after or next.

*An explanation of the nonconvex optimization problem will then **follow**.*

– verb to obey

*We **followed** the program developed in [9] and [15].*

following – noun   that which comes immediately after

*This can be seen in the **following**:*

– adjective   that which will now be described

*...as seen in the **following** table:*

as follows – adverb   what is listed next

*The three theorems are **as follows**:*

hierarchy

Notice that the 'i' comes before the 'e'.

*This must be considered when creating the network ~~heirarchy~~ (**hierarchy**).*

The plural form is 'hierarchies'.

*They extended these ~~hierarchys~~ (**hierarchies**) to...*

The adjective form is 'hierarchical'.

*We plan to build our model based on this **hierarchical** structure.*

in detail

'details' is the plural form of the noun 'detail'.

*The key **details** of this report can be found in the Appendix.*

'In (more) detail' is a fixed phrase meaning 'thoroughly.'

Do not use it in a plural form:

The findings will be discussed ~~in details~~ (**in detail**) in the next section.

indentify

Despite producing many hits on Google, there is actually no such word as 'indentify'.

*We will now **identify** the key reasons for why these faults occurred.*

infer/imply

infer verb – to derive or conclude based on reasoning or evidence

imply verb – to indicate or suggest without actually being stated

These two verbs can be difficult to determine. Although on the surface they represent similar actions, their meanings are quite marked. Use 'infer' when you or someone comes to a conclusion about something based on the evidence available; use 'imply' to suggest an opinion or make an indirect statement that allows the reader to evaluate its value.

*The engineer is able to ~~imply~~ (**infer**) that this is only a temporary interruption.*

*The findings in both [9] and [12] ~~infer~~ (**imply**) that power would be lost immediately.*

Imply can be used for when the writer does not actually believe it to be true.

*This **implies** that the machines can be manufactured within weeks, which is highly unlikely.*

insecure/unsecured

There is debate over both 'insecure' and 'unsecure' but the former is generally preferred (despite the more common social connotation). 'Unsecured' is fine but note the slight difference in meaning between the two accepted terms:

insecure – lacking in security (also 'nonsecure' is gaining popularity)

unsecured – having no security at all

Networks tend to be 'unsecured' and channels 'insecure'.

*A message will appear informing the user that they have joined an **unsecured network**.*

*Yet, [19] proposed a way to establish peer-to-peer authenticated communications over an **insecure** channel.*

inset/insert

An 'inset' is a layer or subwindow that is added to an existing graph in the form of a string or tables of text. You 'insert' that 'inset' into the figure and then that figure is 'inserted' into the text.

instance/instant

instance   noun – a case/example or occurrence

instant   noun or adjective – quick, immediate; a short space of time

These two situations tend to produce errors:

*We observed one ~~instant~~ (**instance**) of this on each router.*

*There was an ~~instance~~ (**instant**) reduction in the calculation time.*

integrate

This verb can be found spelled any number of ways:

*This peripheral interface can ~~intergrate~~ (**integrate**) the flows and direct them accordingly.*

*We ~~integreated~~ (**integrated**) these to facilitate the secure group communication.*

intend/intent

intent/intention noun – purpose; something that is intended.

'intent' is the stronger of the two nouns, with 'intention' meaning a more general purpose or plan for something.

The error occurs when the verb 'to intend' is required instead.

intend verb – to have in mind to do something.

*The attacker does not ~~intent~~ (**intend**) to send malicious traffic currently.*

*This output may be very different to the ~~intented~~ (**intended**) load.*

interconnect

Although primarily a verb, this term also appears as a noun and an adjective in the telecommunications field to mean a connecting device.

verb – *The insulators **interconnect** the loads with the sources of electric power.*

noun – *With a long **interconnect,** this wire spacing is not sufficient for limiting the delay.*

interrupt

Usually a verb but can be used as a noun in engineering to represent a signal that diverts a CPU for prioritizing tasks.

The standard noun of the verb 'to interrupt' is 'interruption'.

*The timer **interrupt** is the principal hardware source for the project.*

*The **interruption** of the fault current is necessary for isolating the line.*

iterate/iterative

Iterate   noun – instructions or loops inserted into a program that repeat until completion.

iteration   noun – the use of repetition in a program.

iterative   adjective – repeating; making repetition.

Iterates can be thought of as the individual parts making up an iteration. The iteration is the process in general and can itself be pluralized if the process takes place multiple times.

*A sequence of **iterates** will then be generated.*

*The weights will be adjusted to reduce the error for the next **iterations.***

*These can easily be obtained by using an **iterative** procedure.*

join/joint

join – verb

joint – adjective/noun

*The ~~join~~ (**joint**) transform correlator (JTC) has been studied in detail [11].*

*This was a ~~join~~ (**joint**) effort by the two organizations.*

*It also indicates that they are willing to ~~joint~~ (**join**) the group.*

keep/remain

remain   verb – to continue to be; to be left; to stay there

keep   verb – to hold or retain; to maintain

'remain' and 'keep' are similar in meaning and either can be used here:

*It is crucial to **keep/remain** calm when this occurs.*

But when the meaning relates to something that continues to be or continues to exist then use 'remain'.

*This can ~~keep~~ (**remain**) a problem for those with poor connections.*

*The rate ~~keeps~~ (**remains**) at 2.45 Mbps in the static situation.*

And when the meaning relates to maintaining or holding on to something use 'keep'.

*They need to ~~remain~~ (**keep**) their address hidden.*

lack/lack of

'of' should not follow the verb 'to lack'.

*The field still ~~lacks of~~ (**lacks**) research examining this inequality.*

'lack' is also a noun and this is often used with 'of'.

There is *a lack of* research dealing with this issue.

A distinction can be made with the comparable verb 'to fail'.

lack   verb – to be without

fail   verb – to fall short in achieving something

*This still ~~lacks~~ (**fails**) to explain why the score was so low on the test.*

lifetime

This is always one word.

*The ~~life time~~ (**lifetime**) of the battery has a large part to play in this.*

linear

linear – adjective

linearity – noun

*This function addresses the ~~nonlinear~~ (**nonlinearity**) of the input variables.*

This is a common spelling error:

*Its application lies in its capability to efficiently model this ~~liner~~ (**linear**) relationship.*

loss/lose

'lose' is the verb and 'loss' is the noun

*This was largely responsible for the ~~lose~~ (**loss**) of power.*

*The impact was apparent on the data ~~loses~~ (**loss**) rate.*

The plural noun is 'losses'.

*This was a direct result of the power ~~loses~~ (**losses**) the previous day.*

When the verb is required be careful not to use the unrelated adjective 'loose'.

*In this case it may ~~loose~~ (**lose**) its neutral earthing.*

maintenance

There is a tendency to use 'maintain' for the first part.

*Routine ~~maintainance~~ (**maintenance**) was not carried out by this company.*

malfunction/dysfunction

dysfunction – not working properly

malfunction – stops working for a period of time; a breakdown

Computers tend to 'malfunction' whereas organs of the body and social concepts and ideas are 'dysfunctional'.

*A ~~dysfunction~~ (**malfunction**) of this nature is labeled 'read upset'.*

manufacture

manufacture – verb

manufacturer/manufacturing – noun

*These transformers are ~~manufactored~~ (**manufactured**) with a third tertiary winding.*

*This has a great impact on the various ~~manufactures~~ (**manufacturers**) of the product.*

misdirect

misdirect – direct to the wrong place or in the wrong direction.

redirect – direct to a new or different place.

*One option we could use is to **misdirect** the attacker.*

redirect can also be a noun:

*This has proven to be a useful **redirect.***

ohmmeter

Although derived from a person's name, do not capitalize the first letter.

*We use an **ohmmeter** to check the resistance.*

Also check the spelling:

*The ~~ohmeter~~ (**ohmmeter**) measures the resistance rather than the current.*

Note that Ohm's law does take a capital letter.

on average

Do not use 'averagely' to mean usually or typically. Use 'on average' instead.

*It takes ~~averagely~~ (**on average**) 0.12 seconds for the network to converge.*

Do not write *~~in average~~*

'averagely' is an adverb meaning moderately or to an average level.

*This has been exacerbated by an **averagely** skilled workforce in the industry.*

on the other hand

Despite making more literal sense, the phrase is actually 'on one hand' not 'in one hand'.

On one hand...........on the other hand

~~In one hand,~~ the infrastructure has been improved recently; ~~in the other hand,~~ the number of system users has declined.

Do not make this careless mistake either:

~~One~~ the other hand...

one by one/one to one

one to one/one on one – direct communication involving two people

one by one – successively; one at a time

*In this subframe there is **one to one** correspondence.*

*The traditional method would process these **one by one**.*

optimum/optimal

optimal – adjective   optimally – adverb

optimum – noun and adjective

A distinction can be made between the two adjectives. 'Optimum' usually describes the ideal or best amount and is used in a more specific sense than 'optimal', which is based on circumstances and relates more to quality.

*This is not an **optimal** solution because the transmittance and reflectance energy levels are too low.*

*We would need to increase this to the **optimum** level.*

Also note that optimal cannot be used as a noun.

*The main objective is of course to find the ~~optimal~~ (**optimum**).*

oscillation

Avoid these common spelling errors:

~~ocillation~~   ~~ocillation~~

*A stabilizing controller was added to improve the dynamic performance and eliminate any **oscillation**.*

output

This is always one word.

*The ~~out put~~ (**output**) of the laser diode is then calculated.*

outage

A rather comical error can occur with this noun.

outage – an interruption or failure in the supply of power

outrage – a feeling of anger

*This can be achieved by maintaining supply under specified ~~outrage~~ (**outage**) conditions [14].*

over-

over time/overtime

The two word term means the passage of time. When referring to extra hours of work it is one word.

*Positive transformations will then take place ~~overtime~~ (**over time**).*

The following are all one word:

~~over head~~   overhead
~~over lapped~~   overlapped
~~over load~~   overload
~~over ride~~   override

pass/past

'passed' is the past tense of the verb 'to pass'

'past' can be a noun, adverb, adjective or preposition. It is not a verb or verb form.

*The systems ~~past~~ (**passed**) these initial tests.*

*This represents the relationship between the **past** load and the current load.*

perceptron/perception

perceptron – an algorithm that computes a single output from multiple real-valued inputs

perception – the act of recognition or understanding

*This is the most common activation function adopted for a multi-layer ~~perception~~ (**perceptron**).*

*It is used to change a viewer's **perception** and works best at low resolution.*

-perform

The following terms are always one word:

underperform   outperform   overperform

*These consistently ~~out perform~~ (**outperform**) the other clustering algorithms.*

Outperform is used comparatively.

Overperform cannot be used comparatively and just means performing better than expected.

*Our system ~~overperformed~~ (**outperformed**) those of [13] and [15].*

periodic/periodically

period – noun
periodic – adjective
periodical – noun/adjective
periodically – adverb

Use 'periodic' for the adjective form as 'periodical' is used primarily as a noun for something that is published periodically.

*This allows them to run in parallel with the main generators for **periodic** load testing.*

permittivity

This tricky term contains three i's and three t's.

*...where μ and ε are magnetic permeability and electric ~~permitivity~~ (**permittivity**) respectively.*

phase

Sometimes errors are caused by their similarity in spoken English.

*This will take place in the design ~~face~~ (**phase**).*

*The work is divided into three sequential ~~faces~~ (**phases**).*

phenomena

The plural of the noun 'phenomenon' is 'phenomena'

*The same ~~phenomenons~~ (**phenomena**) will be expected in the multi-circuit substations.*

*Thermal loss is a ~~phenomena~~ (**phenomenon**) where electrons with excessive energy...*

The plural is not ~~phenomenas~~

point

When using the verb 'to point' to mean 'draw attention to' or 'indicate', make sure you include 'out' and 'to' respectively.

*They also point (**out**) that the rise may not be due to poor circuitry.*

*This trend points (**to**) a possible weak link in the chain.*

precede/proceed

These verbs look similar but have very different meanings.

precede – to come before

proceed – to carry on, continue or advance

*We then **proceeded** to configure the routing protocols.*

*This step **precedes** the traffic being forwarded.*

principal

principal   adjective – first or highest in rank; chief

principle   noun – a rule of action or general law

*The **principal** issues will then be determined.*

*There is a need for a set of **principles** that will provide these values instantly.*

priority

The noun 'priority' does not necessarily need an adjective modifying it.

If this occurred then the original service would be given ~~high~~ priority.

But if an adjective is deemed necessary the following are typical:

high/higher/highest, low/lower/lowest, top, first

*The first step is to disable a running service with ~~small~~ (**low**) priority*

*The packets that are rarest would take ~~larger~~ (**higher**) priority than…*

propagation

Spelling is an issue with the noun form.

~~propergation~~   ~~propogation~~

The verb form is 'to propagate'.

*These techniques limit how far the packet **propagates** across the network.*

*They measured the relationship between ~~propagate~~ (**propagation**) time and receiver power.*

protocol

The ending can sometimes create errors.

*The data security is implemented in the ~~protocall~~ (**protocol**) stack.*

queueing

Check the pattern of e's and u's.

*Various ~~queing~~ (**queueing**) systems have been addressed in ATM networks...*

*The previous ~~qeueuing~~ (**queueing**) model did not include these policies.*

rationale

rational   adjective – reasonable, sensible

rationale   noun – the main reason accounting for something.

The noun 'rationale' is usually required.

*The ~~rational~~ (**rationale**) for this is explained below.*

reduce/reduction

reduce – verb

reduction – noun

'reduce' cannot be used as a noun.

*We were hoping to achieve a power consumption ~~reduce~~ (**reduction**).*

There can be an increase and a decrease but not a reduce – it is a reduction.

rely/reliance

rely – verb

reliance – noun ~~relliance~~

*There is too much ~~rely~~ (**reliance**) on these backup schemes at present.*

The adjective forms are often mixed up.

reliable   adjective – that may be trusted
reliant   adjective – having dependence

*Effective prevention is ~~reliable on~~ (**reliant on**) locating their exact position.*

resonance

resonance (electrical)   noun – the reactance of an inductor balancing the reactance of a capacitor

resonator   noun – a circuit element or system that exhibits resonance

The adjective form is 'resonant'.   *resonant frequency/resonant antenna*

resource

resource – a source of supply or support
recourse – access to something or someone for help or protection

Nine times out of ten the noun required is 'resource'.

*This follows an investigation of the energy ~~recourses~~ (**resources**).*

respectively

'respectively' is used for parallel lists to inform the reader that a second list of things is in the same order as a previous list of things. It is employed to avoid repetition by not having to write out all the elements again.

It is not required if an earlier reference has not been made.

There is a closed dialog box and an open dialog box ~~respectively~~.

This following list of numbers does not refer to any previous list of items, so again respectively is not needed.

The weights chosen were 0.2, 0.4, 0.6 and 0.8 ~~respectively~~.

Here 'respectively' can be used because there is an earlier reference to a list of items:

*Meanwhile, the factors B1, B2 and B3 were assessed and produced weights of 0.18, 0.23 and 0.43 **respectively**.*

respond/response

respond – verb

response – noun

*They will need to ~~response~~ (**respond**) to the attack much quicker.*

*The converter power ~~respond~~ (**response**) is shown in Figure 3.*

rest/remaining

If you are giving a specific number do not use 'rest'.

~~The rest 30~~ cannot be detected at this stage.

The rest cannot be detected at this stage.

You could use:

*The **remaining** 30/The **other** 30 cannot be detected at this stage.*

'rest' is a noun and 'remaining' is an adjective. The following nouns will require 'remaining' to describe them:

*The ~~rest~~ (**remaining**) transformers supply the 33 kV busbar.*

*We will now look at the ~~rest~~ (**remaining**) security goals.*

Do not use the noun 'rest' in the plural form.

*The ~~rests~~ (**rest**) were deemed unnecessary for what we are trying to achieve.*

Whatever 'rest' is referring to will determine whether the verb is singular or plural.

*The rest (of the evidence) **is** considered weak.*

*The rest (of the users) **were** given a different task.*

restore

In computing, 'restore' can be used as a noun to refer to recovering data or a system using a backup measure. Its common use is as a verb.

'restoration' is the traditional noun form and should be used when writing in general terms.

*The **restore** is received at t2.*

*These measures should occur following ~~restore~~ (**restoration**) of power.*

retrieve

retrieve – verb

retrieval – noun

Like 'interrupt' and 'disconnect', retrieve has also developed into a singular countable noun (*a retrieve*) but its primary use is as a verb and the main noun form is retrieval, which is usually uncountable.

This would involve some kind of information ~~retrieve~~ (**retrieval**).

Newly generated content is **retrieved** *by receiver nodes.*

rise/raise

rise   verb – to increase

   noun – an increase; an act of rising

raise   verb – to lift up; to elevate

   noun – an increase in amount

      (especially salary)

'raise' as a verb always has an object linked to it.

Our modification was able to **raise** *the transmission speed in all areas.*

'rise' as a verb is used on its own and does not require a direct object.

The levels will **rise** *again at some point.*

As a rule we 'raise' something but something 'rises'.

'rise' is an irregular verb.

Present simple:

I/we/they   **rise**

it   **rises**

Present participle: **rising**

Past participle: **risen**

Past simple: **rose**

Often the wrong verb or verb form is selected.

*While in 2011, the number of companies producing this data ~~raised~~ (**rose**) to fifteen.*

*The switch had a fast ~~rose~~ (**rise**)/ fall time.*

*They ~~rose~~ (**raised**) the price of each unit to coincide with this.*

robust

robust – adjective

robustness –noun

Consequently, something can be robust or have robustness.

*It is very important that it is ~~robustness~~ (**robust**) in this situation.*

*The main advantages are its ability to utilize the outgoing bandwidth and its **robustness.***

run

The forms of this verb are often confused. The past tense is 'ran' and the third person singular is 'runs'. Do not use 'ran' with the verb to have.

past – it **ran**   past perfect – it had run
we/they ran   we/they had run

present – it **runs**   present perfect – it has run
we/they run   we/they have **run**

future – it will run
we/they will run

*The system will **run** according to these principles.*

*We ~~run~~ (**ran**) it on a simulator and studied the effects of…*

*The network ~~run~~ (**runs**) entirely autonomously.*

*Most of the steps will need to be ~~ran~~ (**run**) separately.*

safeguard

'safeguard' can be a noun or a verb and is always one word.

*This protection scheme is designed to ~~safe guard~~ (**safeguard**) the windings.*

schema/schematic

Schematic can be an adjective but is usually used as a noun to mean a plan or diagram.

*Fig 2.1. **Schematic** of a single-ended 6T SRAM bitcell.*

A schema can have the same meaning but can also refer to an underlying structure or framework of something.

*The **schema** was not suitable for this particular program.*

A scheme is normally used to refer to a program, policy or project.

*The proposed **scheme** certainly shares some characteristics with previous approaches.*

sensor/censor

sensor   noun – a mechanical device that transmits a signal

*Wireless and **sensor** networks will also be explored in more depth.*

censor   noun – someone who assesses and suppresses the work of others

verb – to delete or suppress

*This user can be easily **censored** once their address is known.*

Something without a sensor is 'sensorless' and hopefully not 'senseless'.

*They presented a new senseless (**sensorless**) speed-control scheme.*

sequential

sequence – noun
sequential – adjective

*It should be a known sequential (**sequence**) with good range resolution.*

Check the spelling of the adjective form.

*This can be achieved by transmitting a sequencial (**sequential**) series of individual frequencies.*

setup

As one word 'setup' is a noun meaning arrangement or organization. As two words it is a phrasal verb meaning to begin or make ready for use.

*Fig. 5. Overall ~~set up~~ (**setup**) of the apparatus used.*

*The hardware **setup** includes an LCL circuit and DSP chip.*

*We **set up** this circuit in a similar way.*

shortfall

The noun is always one word.

shortfall   noun – a deficiency or shortage

*There is an estimated annual ~~short fall~~ (**shortfall**) of 32 000 workers in this industry.*

significant/significance

significant – adjective
significance – noun

*There was a ~~significance~~ (**significant**) difference between the two variables.*

*This is statistically ~~significance~~ (**significant**) at the 0.01 level.*

*This chapter will focus on the ~~significant~~ (**significance**) of the scheme.*

simulate/stimulate

simulate – to create a likeness or a model of a system or situation

stimulate – to excite or rouse to action

*We require software that can ~~stimulate~~ (**simulate**) the electromagnetic system.*

spatial

Check the spelling of this adjective.

~~spacial   spatiall~~

*These components must have the same **spatial** volume.*

specificity

'specificity' can be a noun related to the adjective 'specific' or a statistical measure for the number of correctly identified negatives (true negative rate).

*The test produced sensitivity and **specificity** of 67 and 90 percent.*

square

Note the form of the following mathematical measures:

*Least mean **square** algorithms*

*Least **squares** methods (not 'least square')*

*Mean **square** error methods*

*Root mean **square** error*

standalone

The adjective and the noun are one word and the phrasal verb two.

*It was originally designed as a **standalone** program.*

*This would be able to **stand alone** in the absence of any...*

standby

The noun and the adjective are one word.

*They can be found during read operation and in ~~stand by~~ (**standby**) mode.*

stationary/stationery

stationary	adjective – having a fixed position; not moving
stationery	noun – writing materials
stationarity	noun – a process in which the parameters do not change with time.

*We can use both ~~stationery~~ (**stationary**) and mobile devices for this.*

*This can be extended in order to test for **stationarity**.*

step-down

Step-down can be used as both an adjective and a phrasal verb. Only the adjective is hyphenated.

*The **step-down** transformer is used to ~~step-down~~ (**step down**) the voltage from 210…*

stochastic

stochastic   adjective – relating to a process involving a randomly determined sequence of observations

This is frequently misspelled.

~~schotastic~~   ~~stocastic~~   ~~stokastic~~   ~~stochasic~~

*It is a method that relies upon the ~~stocastic~~ (**stochastic**) nature of modern computers.*

Do not confuse with the adjective 'scholastic' – relating to schools or education.

*This engineering award is based solely on **scholastic** ability.*

strategy/strategic

Note the difference between the noun and the adjective forms.

strategy – noun

strategic – adjective

*This is part of the UK's renewable energy ~~strategic~~ (**strategy**).*

*There is an element of ~~strategy~~ (**strategic**) planning in this deployment process.*

strengths and weaknesses

A fairly general term but note the various errors connected with it:

*There are a number of strengths and ~~weakness~~ (**weaknesses**).*

*An evaluation of its strengths and ~~weakens~~ (**weakenesses**) must then follow.*

When used together the two nouns are usually plural unless used in the following way:

This is both a strength and a weakness.

'strength' is a noun that can be either countable or uncountable.

countable – a particular quality or ability.

uncountable – physical or mental power and energy.

A plural is often used in error for the uncountable noun meaning.

These materials are the main ones used in the industry because of their ~~strengths~~ (**strength**).

Also, errors are made with the verb 'to strengthen'.

This ~~strengths~~ (**strengthens**) *the line sufficiently.*

They will also look to ~~strength~~ (**strengthen**) *the links between the two operators.*

survey/surveillance

survey   noun – a sampling; an act to determine the form or position of something

surveillance   noun – continuous observation

A detailed ~~surveillance~~ (**survey**) *has been carried out by [16] and to a lesser extent [12].*

These systems can be enabled for ~~survey~~ (**surveillance**), *failure diagnosis, environmental monitoring...*

synchronous

Spelling is an issue here.

It is useful to adopt a combined ~~synchronus~~ (**synchronous**) *and* ~~asyncronous~~ (**asynchronous**) *approach.*

technique/technical

technique – noun

technical – adjective

*The digital signal processing ~~technical~~ (**technique**) included a threshold filter...*

*An absence of ~~technique~~ (**technical**) data was largely responsible for this.*

test/testify

The verb 'testify' is unrelated to 'test' and means to give evidence, usually in a court of law.

*We will ~~testify~~ (**test**) the validity of this as an initial step.*

throughput

Always one word.

*This new application requires higher ~~through put~~ (**throughput**) than the previous ones.*

totally/in total

totally adverb – in a complete or total manner or degree

Do not use 'totally' to mean 'in total'.

There were sixty users ~~totally~~.

There were ~~totally~~ sixty users.

*There were sixty users **in total**.*

'totally' is an informal word for 'completely' and should be avoided.

This is one of the reasons why the industry is not ~~totally~~ dominated by Western companies.

tradeoff

This can be either hyphenated or written as one word. As a guide, only use the hyphen form when the term is acting as an adjective.

*A **trade-off** inductor value is chosen for the DAB converter.*

The less common phrasal verb is split into two words.

trade off – verb

*This occurs when two suppliers are **trading off** against one another.*

traffic

Traffic is usually described as heavy or light/little. 'High' and 'low' can be used if accompanied by level.

*There are a number of benefits but it does create ~~bigger/higher~~ (**heavier**) network traffic.*

*There would be **little** traffic on the network at this time.*

*A **high** traffic **level** indicates that the modifications have not worked.*

transceiver

Note the following common spelling errors:

*This ~~tranceiver~~ (**transceiver**) will cover distances of up to 80 km.*

*They have developed an optical ~~transciever~~ (**transceiver**) capable of controlling self-heating and expanding the temperature range.*

transient/transience

transient   adjective – lasting only a short time; temporary

The related noun is 'transience'.

'transient' can also be a noun meaning a sudden or brief, and often damaging, increase in current or voltage.

*Some of the line faults were permanent whereas others were **transient**.*

*They determined the duration of the **transient** automatically.*

transmit/transit

transmit   verb – to send or forward to a destination; to emit

transit   noun – the act of passing through; transportation of goods

Transmit is usually the desired term.

*An SFCW radar should ideally ~~transit~~ (**transmit**) all these frequency components.*

troubleshooting

Troubleshooting is always one word.

*Trouble shooting (**Troubleshooting**) the routing issues is important at this stage.*

An actor who carries this out is called a 'troubleshooter'.

trustworthy

There is no such term as 'trustable'.

trust   noun
trustworthy   adjective – dependable, reliable

*The aim is to have links that are highly trustable (**trustworthy**).*

*Similar mechanisms could be used for **trust** anchors here.*

try/attempt

Do not use the noun 'try' with the verb 'to make'. Instead use 'attempt'.

*A try (**An attempt**) was made to fix this problem...*

Ideally, use a verb.

We attempted to fix this problem...

You can give something a try but you 'make an attempt'.

twice/double

Twice can only be used as an adverb; double can be used as a noun, adjective, adverb and verb.

There is some overlap in meaning but as a guideline use 'twice' to mean 'two times' and for comparison alongside 'as'. Use double for expressing quantity, specifically for multiplying by two or as much again in size, strength or number.

*In this scenario the bandwidth would be twice (**doubled**).*

*Recall that [11] recommended carrying this operation out double (**twice**) or even three times.*

*There were double as (**twice as**) many nodes...*

ubiquitous

ubiquitous   adjective – everywhere; all around.

Ubiquitous computing involves embedding processors in everyday objects for communication purposes.

usage

usage   noun – a specific act of using (the manner or the amount)

use   noun – the general act of using

*They noted heavy **usage** on consecutive days.*

*This was probably the earliest **use** of electromagnetic signals to locate…*

validate/validation

validate – verb

validation – noun

*A ~~validate~~ (**validation**) of this information is necessary on these IP networks.*

*In order to **validate** the cable model, a pole-pole DC fault is applied.*

variance/variant

variance – noun

variant – adjective or noun

*These load trends were found to vary between the different inputs, with an average ~~variant~~ (**variance**) of 2.5%.*

*This can be described as a **variant** form of the Elman recurrent network.*

*We noted three **variants** in this particular sample.*

volatile/volatility

volatile – adjective

volatility – noun

*Permanent storage can be classified as ~~volatility~~ (**volatile**) such as main memory or ~~nonvolatility~~ (**nonvolatile**) such as flash memory.*

*Time-varying **volatility** is of great importance in many sectors.*

vulnerable/vulnerability

vulnerable – adjective

vulnerability – noun

On the downside, VSC converters are highly ~~vulnerability~~ (**vulnerable**).

This falsification has led to ~~vulnerable~~ (**vulnerability**) *across the network.*

Spelling can be an issue:

~~vunerable~~   ~~vulnurable~~

warn/alarm

When choosing between these verbs, use 'alarm' for distress or for being startled and 'warn' for caution or a straightforward notification.

The user would then have to be ~~alarmed~~ (**warned**) *that an attack is imminent.*

well/good

good – adjective

well – adverb (can also be an adjective meaning 'healthy')

'good' is mainly used with nouns to describe the subject of the sentence. It comes before the noun.

There was a ~~well~~ (**good**) *signal and this contributed to the positive result.*

In order to form a ~~well~~ (**good**) *understanding, both concepts must be explained.*

So if you are describing a noun use 'good'. 'well' often describes a verb or adjective.

The programmer performed the task ~~good~~ (**well**).

Some of these systems have been **well developed**.

When 'well' is describing the manner of something, it is placed after the word it is modifying.

This ~~well fits~~ *with the scheme and can also be used for...*

This **fits well** *with the scheme and can also be used for...*

*This has proven to **work well**...*

When 'well' modifies an adjective (past participle) it comes before it.

*The method is not particularly **well known**.*

It only ever describes a noun when it is part of a compound.

*However, a **well-designed** method could result in excellent recovery of the input signal.*

widespread

This adjective is always one word.

*A ~~wide spread~~ (**widespread**) installation of MGs could reduce household carbon emissions.*

Take care not to make the following error: ~~wildspread~~

This error also occurs with 'widely used'.

*This is the most ~~wildly~~ (**widely**) used approach in electric load forecasting.*

withstand

Always one word.

*Since each valve must ~~with stand~~ (**withstand**) a high voltage level, these devices need to be in series.*

worth

~~worth to mention~~

The term is 'worth mentioning' because 'worth' is a preposition in these phrases.

*It is ~~worth to mention~~ (**worth mentioning**) that both papers have yet to be published.*

It cannot be used as an adjective to modify a noun.

~~This is a worth mentioning example~~...

*This example is **worth mentioning**...*

The same applies to 'consider' and 'understand'.

*This is worth ~~to consider~~ (**considering**).*

SECTION V

Index

Index

Index